APOCALYPSE 2012

A SCIENTIFIC
INVESTIGATION
INTO
CIVILIZATION'S
END

APOCALYPSE
2012

LAWRENCE E. JOSEPH

MORGAN ROAD BOOKS
New York

MORGAN ROAD BOOKS

PUBLISHED BY MORGAN ROAD BOOKS

Published in the United States by Morgan Road Books, an imprint of The Doubleday Broadway Publishing Group, a division of Random House, Inc., New York.
www.morganroadbooks.com

MORGAN ROAD BOOKS and the M colophon are trademarks of Random House, Inc.

Book design by Lee Fukui

Library of Congress Cataloging-in-Publication Data

Joseph, Lawrence E.
    Apocalypse 2012 : a scientific investigation into civilization's end / Lawrence E. Joseph.—
1st ed.
        p.   cm.
    Includes bibliographical references and index.
    1. Two thousand twelve, A.D. 2. Twenty-first century—Forecasts. 3. Catastrophical, The.
4. Science and civilization. I. Title.

CB161.J67 2006
303.4909'05—dc22

                                                                                2006049847

ISBN: 978-0-7679-2447-4

PRINTED IN THE UNITED STATES OF AMERICA

10 9 8 7 6 5 4 3 2 1

First Edition

To Phoebe and Milo. I love you.

CONTENTS

Acknowledgments   ix

INTRODUCTION   1

GUILTY OF APOCALYPSE:
THE CASE AGAINST 2012   16

SECTION I: TIME

1. WHY 2012, EXACTLY?   23

2. THE SERPENT AND THE JAGUAR   34

SECTION II: EARTH

3. THE MAW OF 2012   47

4. HELLFIRES BURNING   58

5. CROSSING ATITLÁN   73

SECTION III: **SUN**

    6. SEE SUN. SEE SUN SPOT.   87

    7. AFRICA CRACKING, EUROPE NEXT   102

SECTION IV: **SPACE**

    8. HEADING INTO THE ENERGY CLOUD   119

    9. THROUGH THE THINKING GLASS   132

SECTION V: **EXTINCTION**

    10. OOF!   155

SECTION VI: **ARMAGEDDON**

    11. LET THE END-TIMES ROLL   171

    12. HAIL THE STATUS QUO   188

    13. 2012, THE STRANGE ATTRACTOR   199

    CONCLUSION   219

    Notes   233

    References   241

    Index   251

ACKNOWLEDGMENTS

As this book considers the possibility of life as we know it changing radically in the very near future, I would be remiss not to acknowledge Whomever or whatever gave us all that we hold so dear. Whether you are God, as I believe, or Gaia, the Big Bang, or some other entity or concept entirely, thank you for all the joy, excitement, pride, wonder, and love. And thanks for all the bad stuff, too. Existence is better than void.

Many people helped me with this book and deserve my gratitude. Carlos and Gerardo Barrios, Mayan shamans from Guatemala, shared their wisdom and insight and opened many doors. Thanks also to the Saq Be Institute of Santa Fe, New Mexico, which introduced me to the Barrios brothers. Anne Stander, of Johannesburg, South Africa, opened doors of perception and became a good friend.

In Akademgorodok, Siberia, Alexey Dmitriev enlightened me regarding the interstellar milieu. Alexander Trofimov of ISRICA (International Scientific-Research Institute of Cosmic Anthropoecology) helped me understand time as part energy flow, part dimension. Thanks also to Olga Luckashenko, the brilliant interpreter who guided me through some very dense moments.

Though I did not work with them on this book, James Lovelock, of Cornwall, England, and Lynn Margulis, of Amherst, Massachusetts, are always inspiring. David A. Weiss, of Packaged Facts Inc. of New York City, taught me more than I ever realized about managing data. Not quite sure whom to thank for the Internet, but without it, this book could never have taken shape.

John and Andrea, Scott and Terry, Chris, Brent, Jack, Jay, Larry and Marilyn, Ed and Carol, Mitch, Frances and Hans Eric are great folks. Hail, Mirabitos. Susan and company have been a blessing to my family. Arthur and Jessica, too. Erica, absolutely. Dr. Jon, Jason, and Jose are fine new friends. And Sherry, there's no one quite like you.

A lifetime of thanks to my mother, Yvonne Joseph, who is perfectly encouraging. Her mother, Hasiba Shehab Haddad, told me the family stories that have given context and trajectory to my life. And Dad, I still feel your warmth.

Andrew Stuart, my literary agent, has great instincts, proper manners, and brass balls.

Amy Hertz, the editor and publisher of this book, is a blast. Every comment she made improved the book, and she made a lot of comments, ranging from tart admonishments about grammar, syntax, and clarity to gently advising that I "shout at the top of my lungs, then go have a good cry," the latter not being something we Brooklyn boys usually do. Amy, you are one of a kind. Working with you was the experience of a lifetime.

INTRODUCTION

On the first day of freshman writing class, the instructor told us that good writing was all about emotions—portraying them, eliciting them, unraveling them, being true to them. I stuck up my hand and stammered out something to the effect that, to me, emotions were just the details, and that what really mattered was whether or not people got to stay alive in order to have any. Happy, sad, angry, diffident, deep or shallow, shared with a loved one or burning from within—that's all very interesting, but of secondary importance compared, say, to whether or not one is poisoned to death, or burnt to a crisp.

So when I first heard about how the world might end in 2012, I took to the idea right away. Except that no one in his right mind believes the world is really going to end. That's the kind of thing weird men wearing sandwich boards and giving out smudgy pamphlets with lots of exclamation points on them like to claim. Theoretically, of course, the world must burn, freeze, crumble, or existentially wig out one day, but that's billions of years down the road, right? Who knows, maybe by then we'll all have moved to another planet, or even figured out a cure for time. But for all practical purposes, the unfathomable concept of the world coming to an end is used mostly to put things in perspective, as in "it's not the end of the world" if your pants don't get back from the dry cleaners until Monday.

There are any number of end-time scenarios, from Hitler/bin Laden/Pol Pot getting his finger on the button, to an asteroid the size of Everest cracking the Earth like an apple, to the Lord God Almighty saying enough is enough. But our planet does not have to literally disintegrate, or all its inhabitants perish, for our world to come to an end, or close enough. If civilization as we know it, that burgeoning and magnificent social, political, and cultural entity, were damaged to the point where its evolution was retarded, where normal relations between nations were disrupted, where a significant percentage of human beings lost their lives and most of the rest faced a future of hardship and horror—that would count.

Since the early 1990s, I have been involved with a company that has sought to help save the world from poisoning itself. Aerospace Consulting Corporation (AC2), of which I am currently chairman, has begged, borrowed, and blood-from-stoned about $10 million to develop the Vulcan Plasma Disintegrator, U.S. patent #7,026,570 B2, a portable, ultra-high-temperature furnace that will completely dissociate highly toxic wastes, including but not limited to lethal biological and chemical weapons that cannot otherwise be disposed of. The Vulcan, when it is finally produced, will be a fifty-yard tube with a robotic arm sticking out at one end. The arm grasps a fifty-five-gallon drum of hazardous, nonnuclear waste, samples its contents to prepare the right settings, sticks it inside the tube, which then heats up to 10,000 degrees, and zaps that sucker, container and all, into nothing: zero toxic residue.

There was always plenty of office space available at the Inhalation Toxicology Laboratory, out on Kirtland Air Force Base in Albuquerque, New Mexico. For next to nothing, our company had a nice suite and complimentary coffee station in the building out behind the kennel with the hundred identical dogs. True, the commute was an ordeal. After going through various security checkpoints, you had to drive all the way around the Electromagnetic Pulse (EMP) Testing Center, a giant wooden platform held together without a single metal nail or screw, on which they would zap, say, a specially shielded 747 jumbo jet, to see if its instruments would fry. Next was the Big Melt Laser Laboratory; no one would ever tell me what it was they melted. Then mile after mile of intercontinental ballistic missiles (ICBMs) in their silos, dug into the hillside. The temptation to speed past them all had to be resisted because that part of the base is shoot-to-kill for vehicles violating the 30-mile-per-hour speed limit or any of the other traffic laws.

Over the past five years we have received considerable support and encouragement from Kirtland Air Force Base, a Department of Defense facility, and from Sandia National Laboratories, a Department of Energy facility responsible for, among other things, the construction and maintenance of every nuclear warhead in the United States.

For the record, neither AC2, Kirtland Air Force Base, nor Sandia National Laboratories, nor any employees or contractual workers associated with those entities are known to take any position whatsoever on predictions concerning the year 2012.

You don't need dire predictions about Apocalypse 2012 to freak out a little about all the weird stuff we've invented that could destroy the world. More than enough biochemical weapons are stockpiled around the globe, starting with mustard gas, a deadly paralytic agent left over from World War I, on through anthrax, sarin, and a variety of other classified compounds, to keep the Vulcan incinerating for many years to come. And the good news/bad news is that there will be even more incredibly toxic stuff to burn up in the future, at least according to those who share the fears voiced by Stephen Hawking, who believes that humankind will extinguish itself from the face of the planet through the misuse of biological weapons:

"I don't think the human race will survive the next thousand years unless we spread into space. There are too many accidents that can befall life on a single planet," Hawking told Britain's *Daily Telegraph*. Hawking, the Lucasian Professor of Mathematics at the University of Cambridge, expressed the opinion that the threat was not so much from a Cold War–style nuclear holocaust as from a more insidious form. "In the long term, I am more worried about biology. Nuclear weapons need large facilities, but genetic engineering can be done in a small lab."

What manner of vile pestilence will renegade eggheads concoct with their gene splicers? They might try to "improve" upon the worst Nature has to offer. For example, some of the latest strains of superbacteria have an enzyme called VIM-2 that breaks down antibiotics. Genetically enhancing the VIM-2 enzyme could give the resulting superorganism a head start so big that antibiotics could never catch up. Perhaps the gene-splicing sociopaths will create "priobots." By bolstering the already formidable self-replicating abilities of prions, these new predatory proteins could turn our brains into useless sponges through Creutzfeldt-Jakob disease, also known as mad cow disease. The priobots might also cause an epidemic of kuru, a brain disorder in which cannibals have been known to giggle themselves to death. How's that for an evil genius's last laugh?

Even if we catch these malefactors before they can do harm, the poisons that they cook up will have to be disposed of. But there's no furnace hot enough to burn up such compounds without leaving toxic residue. That's the niche that Vulcan seeks to fill. It just might save the world after all. That is, as long as it doesn't explode. Since it's planned as the hottest furnace in the world and filled with deadly waste materials, we've had to make damn sure the device is stable and secure. In fact, Vulcan's underlying plasma containment technology has potential applications as a rocket thruster: basically, you just take one end off the containment tube, and zoom, the unit takes off. Upon command, presumably.

ATOM SMASHING

Running a Vulcan furnace requires a megawatt of direct electrical power, enough to run about 25 contemporary, standard, three-bedroom homes, or 200 rent-controlled apartments in Park Slope, Brooklyn, where Victor Simuoli and I planned to construct our atom smasher for the annual Junior High School 51 science fair. The Atomic Energy Commission had kindly sent us the plans for a linear accelerator, a device that propels subatomic particles from either end toward the middle and then smashes them into each other head-on at terrific speeds. Seeing how incredibly complicated the blueprints were, and how running the atom smasher would probably have shorted out the whole neighborhood, Victor and I settled, as I recall, for making a crystal radio receiver out of a cigar box.

We probably wouldn't have given up so easily if we knew there was a possibility that our atom smasher could potentially create a tiny black hole that would eventually destroy the world. Not, mind you, that we were pre-

Columbine or anything, just that, as two nerdy adolescents, the temptation of unleashing *Star Trek*–scale forces would have been hard to resist.

Though our machine would have been way too small to punch a black hole into space-time, the same cannot be said for the large hadron collider (LHC), a 27-kilometer circle on the border between France and Switzerland. When it begins operation in 2007, it will pack the colossal wallop of 14 trillion electron volts. A trillion electron volts, it turns out, is about the same amount of energy used by a mosquito to fly. The remarkable thing about the LHC is that it will concentrate its energy beam into a space one-trillionth the size of a mosquito, smashing protons into 10,000 pieces or more.

According to physicist Michio Kaku, the LHC's incredible focusing power will create "an entire zoo of subatomic particles not seen since the Big Bang," including mini black holes. Mini black holes? Intellectually scintillating though such a smash-up may be, questions must be raised about the calamity potential for some of these experiments. Don't black holes, mini and otherwise, have a tendency to suck up everything around them into oblivion?

Martin Rees, a colleague of Hawking's at the University of Cambridge, is a physicist who also has the distinction of serving as the United Kingdom's Royal Astronomer. Rees warns that the shower of quarks resulting from proton-antiproton collisions could create mini black holes, called strangelets, which have the capability of contagiously converting everything they encounter into a new, hyperdense form of matter. Atoms are made mostly of empty space, space that would be squeezed out by the strangelet, compressing the Earth into an inert sphere about the size of a Home Depot.

An inglorious ending, that.

GRAY GOO

There's always a risk of unanticipated outcomes with new inventions—for example, the "gray goo" scenario that they try not to talk much about, up the road at Los Alamos National Laboratory, famed as the atom bomb's birthplace. Los Alamos is a leader in nanotechnology, which seeks to create nanoscale (billionth of a meter) machines designed to behave like the ribosomes in the cells of our body, assembling complex structures, such as proteins, out of simpler compounds, such as nitrogen, a key component. Nanotechnologists have discovered that, given the right circumstances, the atoms of certain elements naturally assemble themselves into complex structures; germanium atoms will, like cheerleaders at a football game, climb on top of each other to form a pyramid, defying the natural tendency of most atoms, and most noncheerleaders, to give in to gravity and remain on the ground. This self-assembly property proves quite convenient for all sorts of nanoscale endeavors, from breeding ultrapowerful computer chips from bacteria to creating infinitesimal machines that can be injected into the bloodstream to eat up cancers or infections.

What if the nanomachines' appetites got out of control? The result would be gray goo, a term coined by nanotechnology pioneer Eric Drexler, in *Engines of Creation*. Gray goo is a hypothetical nanosubstance that keeps on reproducing itself until it devours all the carbon, hydrogen, and whatever other elements it lusts for and has gooed over the face of the Earth. Imagine the parts of a box of Tinkertoys, carefully laid out on the right kind of mat, assembling themselves into, say, a Tinkertoy robot. Kind of cool. But now imagine that process going haywire, Tinkertoy Robot #1 making Tinkertoy Robot #2, and then those two making two more, and then those four making four more, with the numbers doubling into the thousands, millions, billions, in a runaway process that would continue until the world's raw materials were consumed.

According to Drexler, rapidly self-replicating nanomachines could outweigh the Earth in less than two days. The good news is that something would undoubtedly come along to devour the gray goo. The bad news of course is that there would then be untold gray goo devourers to deal with.

Save the world. Destroy the world. It's all pretty much the same ambition—that is, to prove beyond a shadow of a doubt the absurd proposition that life is divided into two roughly equal halves: oneself, and the rest of the Universe.

Strangelet black holes Pac-Manning the Earth. Priobots infesting our brains. Gray goo engulfing Life as we know it. Way too weird to lose any sleep over, but a man can dream. Ever since kindergarten days, when I and Marty Raichalk would spend hours in the backyard of the house our families shared on a dirt road in Danbury, Connecticut, protecting our imaginary girlfriends, Betty and Sue, from crazed murderers and bumblebees, I'd been waiting, you know, for an opportunity to show off my skills. In grade school

I ached to take on the Martians plotting to steal my brain. One's valuables must be safeguarded, for the sake of all decent people. And who knows how much evil could have been vaporized if Victor and I had ever managed to plug in that atom smasher?

Not to say that Vulcan, if we ever really get it up and running, won't prove lucrative. But that's just milkshake, and what we're talking here is pure malted ego, so rich you can sip yourself into a coma. To save the world from poisoning itself, the planet and the people. Now that would be taking a bow.

SOLAR INDIGESTION

Being of Lebanese descent and therefore somewhat dark-skinned, I've always had a rather arrogant attitude toward the Sun—problems associated with it were what white folks had to worry about. So I couldn't be bothered at first, when Roger Remy, our company's principal scientist and founder, announced that the Sun was "making mayonnaise," which in his idiosyncratic vernacular means "having a breakdown." Roger is kind of a French Moroccan Indiana Jones gone-to-seed, who talks a lot about covert operations, known as "skunk projects," and space travel. But his specialty is the manipulation of plasmas, intensely hot ionized gases, of which the Sun is an immense ball, so I couldn't just dismiss his statement outright.

Whatever the Sun's problems, they were 93 million miles away, unlike Christmas, which at the time, November 2004, was bearing down like a freight train. So with two young children, an exhausted wife, and overbooked holiday travel plans, I let the matter drop.

"The Sun can't get sick, you silly," said my four-year-old daughter, Phoebe, who must have overheard a conversation. I was happy to agree.

On the day after Christmas, a close family friend died of an overdose of narcotics and antidepressants. The overdose was intentional, but the resulting suicide apparently was not. That day, December 26, 2004, was also the day the tsunami struck the Indian Ocean. In the week that followed, my wife grew more distraught over the death of her friend, a young woman of eighteen whom my wife had known since the girl's infancy, while I became preoccupied with the aftermath of the tsunami. I am sorry to say that neither of us had much compassion for the other's grief. The photo I will never forget, on the front page of the *New York Times*, was of a dozen or so people on a beautiful beach—Phuket, Thailand, as I recall—watching the unimaginable wave bear down on them. They looked so defenseless in their skimpy bathing suits. A few were running, but most were just slackjaw transfixed. All died, most likely. Why I felt more for a few figures in a photograph than my wife's young friend, and why my wife felt more for the loss of one troubled teenager than 250,000 people in eleven nations, cannot be explained, except that we're different.

Although the connection between the behavior of the Sun and the Indian Ocean tsunami is debatable, the sheer magnitude of that disaster, so out of the blue, made checking out Roger's mayonnaise hypothesis seem the prudent thing to do. So after the holidays I looked into the matter and, sure enough, the Sun seemed like it had eaten some bad mayonnaise. It was mottled with sunspots, which are larger-than-Earth magnetic storms that can unleash as much energy as 10 billion hydrogen bombs, according to Tony Phillips, editor of the excellent Web site science.nasa.gov. The sunspots were belching billion-ton proton blasts and trillion-volt electron skewers all around the Solar System. Very dramatic, but isn't this how the Sun normally conducts itself?

Not really. Ever since Galileo invented the telescope in 1610, solar activity has been observed to follow cycles of roughly eleven years, activity being judged by the number of sunspots popping up. When I started my research in January 2005, the sunspot cycle was, by scientific consensus, approaching the solar minimum, that is, the period of lowest solar activity, which bottomed out in 2006. Instead, for some unknown reason, the Sun has been throwing a tantrum ever since Halloween 2003, when the largest radiation storms ever recorded pounded the Solar System. Thank goodness most of the Halloween outbursts happened to miss the Earth; they were about twice as strong as the March 1989 solar radiation storm that popped the Hydro-Quebec power grid, blacking out the households of 6 million unsuspecting Canadians. Solar activity remained abnormally high and spiked with the giant sunspots of January 20, 2005, which pelted the Earth's atmosphere with its largest proton storm in fifteen years. What made this all the more astonishing is that it occurred at or near solar minimum, the point in the eleven-year sunspot cycle where there is supposed to be little or no solar activity. Chilling, but not nearly as chilling as September, when the Sun went from perfectly calm, not a blemish, to being covered with sunspots and spitting out record-setting mouthfuls of radiation, right at the height of the hurricane season that produced Katrina, Rita, Wilma, and so many others.

There is nothing in the human experience, including the sacred concept of Almighty God, as reliable as the Sun. The Sun empowers Earthly life. It warms the land and the oceans, begets all plant and animal growth, energizes the atmosphere, helps generate clouds, drives the wind and the ocean currents, and cycles the planet's water supply. The notion, therefore, that the Sun may somehow be changing in any way is the very definition of unthinkable—far beyond the leap required, for example, to grasp the consequences of all-out nuclear holocaust, as Herman Kahn and other doomsday philosophers were once wont to do.

An increase of as little as 0.5 percent in the Sun's energy output would be enough to fry the satellite system on which global telecommunications, military security, and banking depend. Ditto our skin, with spikes in cancer and other radiation diseases. Runaway global warming, and the attendant upsurge in sea levels and flooding, megastorm activity, and even seismic and volcanic holocausts, would seem inevitable.

Having reported on science and nature for more than twenty years, I expected roadblocks in researching this bizarre solar behavior. Famous institutions would naturally be loath to associate their names with such a potentially devastating subject as the changing Sun, for the very good reason that their stamp of authority might cause panic in certain quarters. So I was taken aback to find that the Max Planck Institute, Germany's equivalent of MIT and CalTech, has conducted a number of studies confirming that the Sun hasn't been this turbulent for 11,000 years at least. Ever since the 1940s, and in particular since 2003, solar activity levels have been off the charts. We could be zapped at any moment.

COLLATERAL DAMAGE

Maybe the most frightening apocalypse scenario of all is what's happening in space. Talk about change being unthinkable. I mean, besides a few asteroids here and there, space is just there, right? It doesn't change. Well, the whole Solar System is becoming increasingly agitated because we are moving into an interstellar energy cloud, according to an emerging Russian school of planetary geophysics. These scientists, who base their findings on decades of analyzing satellite data, have found that all the planets' atmospheres, including the Earth's, are beginning to show the effects of this massive input, both directly from the energy cloud and indirectly from the disturbances being created within the Sun from its encounter with the energy cloud.

Not to worry. The Earth's atmosphere will protect us, right? Maybe in the old days it would have, but now Harvard and NASA scientists are reporting that California-sized cracks have inexplicably opened up in the Earth's magnetic field, our essential shield against solar radiation and the deadly cancers and climatic disturbances that come with it. Some scientists are even predicting that a pole reversal, in which the North and South magnetic poles switch places, is imminent. That's a several-thousand-year process in which multiple magnetic pole sites pop up around the globe, confusing and sometimes extinguishing the thousands of species of birds, fish, and mammals that depend upon magnetism for their sense of direction. During the confusion, the Earth's magnetic protection drops to near zero, the cosmic equivalent of a very pale person getting caught on a beach in Miami with no hat, no shade, no sunblock, and an imperfect ass in a teeny Speedo.

One source of protection from excess solar radiation comes from another way the world might end. The sky could fill up with ray-absorbent ash, but that's about the only good news I could find in a BBC documentary reporting that Yellowstone, probably the largest supervolcano in the world, is preparing to erupt. The last time Yellowstone erupted, 600,000 years ago, it vomited enough dust to cover the North American continent several feet deep. Today such an eruption would lead to a nuclear-winter-type scenario that would savage global agriculture and economy, killing hundreds of millions.

And the biggest reason to worry about the end of life is the prediction in *Nature,* perhaps the world's most respected science journal, that at least three-quarters of the Earth's species are wiped out every 62 to 65 million years. It has been 65 million years since the Cretaceous-Tertiary disaster extinguished the dinosaurs, meaning that we are now overdue for a cataclysm that will without doubt reduce our population by at least half, smash our infrastructure to smithereens, and drive most of whatever is left of our civilization underground.

If Yellowstone blows or the Sun's acne festers into boils, ecological problems like ozone holes and global warming will be fondly lamented, the way we started out the 1980s worrying about herpes simplex and ended up with the scourge of AIDS. But the good news, as the irrepressible Admiral Hyman Rickover liked to point out, is that, whatever happens, "a new and wiser species will evolve."

DOWN THE RABBIT HOLE

Firm dates are hard to come by in the disaster prediction game, and about the only thing scientists seemed to agree upon is that whatever was happening now, as we approached the solar minimum, would pale in comparison to the unprecedented turbulence projected for the next solar maximum, expected in 2012.

On impulse, I googled "2012" and promptly fell down the rabbit hole into a thriving apocalypse subculture. Blogs, books, music, and art from every continent prophesied doom for that year. Exponents of a bewildering array of ideologies and philosophies, from indigenous cultures, the Bible, the I Ching, point to 2012 as the time of Apocalypse. Could it be just a coincidence? Or is it more reasonable to assume that divinely inspired traditions would, after all, reach congruent conclusions about the fate of humanity?

"Twenty twelve! That's when, you know, it's all supposed to happen. Big time!" exclaimed our nanny, Erica, when I mentioned my discovery the next morning. A bowl of popcorn would have emptied fast as Erica, a late-night net surfer and talk-radio devotee, burbled with dire predictions and assurances that 2012 is the real Y2K. She seemed to see it all as kind of an ongoing reality show, of the horror variety. Several of her friends were into this doomsday 2012 thing as well, and she gaily recounted some of their suggestions for what to do as The End draws near: "Pass the bong. Build a spaceship. Move underground. Have lots of sex. Commit suicide. See the world. Go about your business. Stop taking your medication. Start taking someone else's. Write that novel. Euthanize your family. Hit Vegas. Praise Allah. Take revenge. Take a crash course in astral projection. Be sure to get a good seat for the ultimate fireworks display."

Why the year 2012, specifically? The hubbub had nothing to do with that being the projected date of the next solar maximum in the sunspot cycle. In fact, there was little or no mention of the Sun, or for that matter science topics in general, among those prophesying doom. Galvanizing the movement was an utterly ancient prediction from Mayan mythology that Time will either end or begin on the winter solstice, December 21, 2012.

At that point I almost dropped the whole thing, because, how to put this . . . I am not New Agey. I am your basic Brooklyn wiseguy Beeming around Beverly Hills. Not that all that ancient oojie-boojie is necessarily invalid, just that most of it is lost on me.

THE MAYAN PROPHECIES

Ancient Mayan astronomy is anything but oojie-boojie. It is a staggering intellectual achievement, equivalent in magnitude to ancient Egyptian geometry or to Greek philosophy. Without telescopes or any other apparatus, Mayan astronomers calculated the length of the lunar month to be 29.53020 days, within 34 seconds of what we now know to be its actual length of 29.53059 days. Overall, the two-thousand-year-old Mayan calendar is believed by many to be more accurate than the five-hundred-year-old Gregorian calendar we use today.

The Maya were obsessed with time. Over the centuries, they devised at least twenty calendars, attuned to the cycles of everything from pregnancy to the harvest, from the Moon to Venus, whose orbit they calculated accurately to 1 day every 1,000 years. After centuries of observations, their astronomers came to the conclusion that on the winter solstice of 2012, 12/21/12, or 13.0.0.0.0 by what is known as their Long Count calendar, a new era in human history will commence. This 12/21/12 "stroke of midnight" begins a new age, just as the Earth's completion of its orbit around the Sun brings a new year at the stroke of midnight every January 1. But so what? Aside from a change in date and a day off from work, there is no inherent, palpable difference between December 31 and January 1—it's not as though we go from cold and dark one day to warm and sunny the next. For that matter, there is no inherent, palpable difference between one year and the next, unless such difference is externally ascribed: going from 1999 to 2000, Y2K was nothing but a transition from a digitally unremarkable number to a nice big round one. It proved to be about as spiritually resonant as an odometer change.

The date 12/21/12 has significance beyond numerical happenstance. It is the annual winter solstice, when the Northern Hemisphere is farthest away from the Sun, and when therefore there is the least daylight and the longest night. On that date our Solar System will eclipse—interpose itself so as to block the view from Earth—the center of the Milky Way. The dark hole at the center of the galaxy spiral was considered the Milky Way's womb by the ancients and now also by contemporary astronomers, who believe that that's the spot where our galaxy's stars are created. Indeed, there's a vast black hole right at the center, making for a nice navel motif.

The Mayan ancients held that 12/21/12 would begin a new age, in vital fact as well as calendar technicality. The date thus portends a most sacred, propitious, and dangerous moment in our history, destined, they believed, to bring both catastrophe and revelation. The years leading up to it presage this awesome potential in terrible and wonderful ways.

I went to Guatemala to evaluate the beliefs and predictions attached to 12/21/12 and concluded, in a nutshell, that the Maya have a track record that is impossible to ignore. Always give genius the benefit of the doubt, and the ancient Mayan astronomers were indeed geniuses. The Mayan prophecies concerning 2012 seem therefore to contain wisdom not necessarily beyond science, but most likely beyond anything contemporary scientific methodology could prove, or disprove, in the short time remaining before the apocalypse deadline.

What possessed the Maya to devote so much exquisite work to astronomy, while never even getting around, for example, to inventing the wheel or even simple metal tools, I cannot say. But simply to ignore their fundamental conclusion that December 21, 2012, is a pivotal date in human history—especially given the profoundly disturbing set of concurrences regarding the 2012 deadline in fields ranging from solar physics to Eastern philosophy—would be foolish in the extreme.

DISCLAIMERS

Some disclaimers are in order here:

I represent no religious or political ideology nor have I, to the very best of my knowledge, fallen under the influence of any individual or group with views relating to 2012. Unlike many of those concerned with end-times, Apocalypse, or Armageddon, I have had no divine revelations, no instructions from alien intelligence, no channelings from ancient sages, no numerological epiphanies.

Neither am I one of those skeptical balloon-puncturers who deflate every notion not 100 percent supported by available physical evidence. Lord save us from the dearth of artistry and creativity that would inevitably result were those killjoys ever to gain the power that they think logic dictates should be theirs.

Nor am I a catastrophe buff. I am proud to report that I expended not one cent or one minute defending against the possibility of the Y2K computer bug. Neither have I ever prepared myself nor my household for nuclear holocaust, comet impact, harmonic convergence gone haywire, or any other such donnybrook. Living in the earthquake zone of Southern California, I do however keep a flashlight by the bed and an extra jug of water in the closet. And for the record, I do not hope, advocate, agitate, or pray for any catastrophe, 2012-related or otherwise, regardless of how uplifting the outcome is purported to be.

My conclusions concerning the potentially cataclysmic nature of 2012 are based on approximately fifteen months of research, conducted with the expertise gained from more than twenty years as an author of nonfiction books and as a journalist covering science, nature, religion, and politics for a variety of publications, most frequently the *New York Times*.

Is writing this book an irresponsible thing to do, for fear of the panic it might cause? The public's right to know is not absolute, but neither is it contingent upon the paternalistic assessments of the global oligarchy. I can only have faith in the overall process wherein the powers-that-be, using their best judgment, attempt to control information that might cause social instability, and also where passionate individuals, groups, and organizations work to bring vital facts to light. Ultimately, the best solutions come from a spirited interchange between truth-seeking individuals and the power structures created to protect us.

THE MARK OF DESTINY

Will the world end in 2012? Will all hell break loose, on the order of an all-out, World War III–scale nuclear holocaust or a meteoric impact like the one believed to have extinguished the dinosaurs? I do not believe so, though that may be partly a reflection of my emotional limitations—as the father of two young, wonderful children, I am simply not capable of such a conviction. Not capable of confronting the possibility that everyone and everything that anyone has ever held dear could be destroyed.

What I am capable of doing is gathering the facts and presenting the evidence necessary to ferret out the reality of 2012. I have found that the prospect of an apocalypse in 2012 should be treated with respect and fear.

This book will demonstrate what I consider the middle-case scenario, namely that 2012 is destined to be a year of unprecedented turmoil and upheaval. Whether the birth agony of a New Age or simply the death throes of our current era, a disturbing confluence of scientific, religious, and historical trends indicates that an onslaught of disasters and revelations, man-made, natural, and quite possibly supernatural, will culminate tumultuously.

The year 2012 has the mark of destiny upon it. Judging from the facts gathered for this book, there is at least an even chance of some massive tragedy and/or great awakening occurring or commencing in that year. The question ultimately is not if but when, not so much the exact date as whether or not this transformational event will occur within our own or our loved ones' lifetime. The value of the 2012 deadline is that, being so close, it forces us to confront the myriad possibilities for global catastrophe, to gauge their likelihood and destructive potential, and to examine how prepared we are to respond to them, individually and as a civilization.

Everyone responds to deadlines, constructively or otherwise. Especially if there's pressure. It's human nature. The last two minutes of each half of a football game, together less than 7 percent of the total playing time, yield at least half the action. I need deadlines. Most of us do. With the unlikely exception of Y2K, that silly dress rehearsal, the 2012 deadline is the first in modern history when so much is on the line for so many.

The blessing of a deadline is the advance notice that goes with it, to get body, mind, and soul together, to take some sensible precautions for oneself and one's family. In some sense, not necessarily including physical survival, we've all got a chance like never before to come together and rise to our collective higher Self. That's the invigorating challenge of 2012. It forces us to find a common purpose. And having a purpose in life is about the surest way I know to stave off demise.

The thesis of this book is that the year 2012 will be pivotal, perhaps catastrophic, possibly revelatory, to a degree unmatched in human history.

1. Ancient Mayan prophecies based on two millennia of meticulous astronomical observations indicate that 12/21/12 will mark the birth of a new age, accompanied, as all births are, by blood and agony as well as hope and promise.

2. Since the 1940s, and particularly since 2003, the Sun has behaved more tumultuously than any time since the rapid global warming that accompanied the melting of the last Ice Age 11,000 years ago. Solar physicists concur that solar activity will next peak, at record-setting levels, in 2012.

3. Storms on the Sun are related to storms on the Earth. The great wave of 2005 hurricanes Katrina, Rita, and Wilma coincided with one of the stormiest weeks in the recorded history of the Sun.

4. The Earth's magnetic field, our primary defense against harmful solar radiation, has begun to dwindle, with California-sized cracks opening up randomly. A pole shift, in which such protection falls nearly to zero as the North and South magnetic poles reverse position, may well be under way.

5. Russian geophysicists believe that the Solar System has entered an interstellar energy cloud. This cloud is energizing and destabilizing the Sun and all the planets' atmospheres. Their predictions for catastrophe resulting from the Earth's encounter with this energy cloud range from 2010 to 2020.

6. Physicists at UC Berkeley, who discovered that the dinosaurs and 70 percent of all other species on Earth were extinguished by the impact of a comet or asteroid 65 million years ago, maintain with 99 percent certainty, that we are now overdue for another such megacatastrophe.

7. The Yellowstone supervolcano, which erupts catastrophically every 600,000 to 700,000 years, is preparing to blow. The most recent eruption of comparable magnitude, at Lake Toba, Indonesia, 74,000 years ago, led to the death of more than 90 percent of the world's population at the time.

8. Eastern philosophies, such as the I Ching, The Chinese Book of Changes, and Hindu theology, have been plausibly interpreted as supporting the 2012 end date, as have a range of indigenous belief systems.

9. At least one scholarly interpretation of the Bible predicts that the Earth will be annihilated in 2012. The burgeoning Armageddonist movement of Muslims, Christians, and Jews actively seeks to precipitate the final end-times battle.

10. Have a nice day.

TIME

That the Rastafarian cabdriver sang reggae prayers to the Almighty Jah all the way to the airport, bowing his head right down to the steering wheel at least fifty times while shooting the rapids of I405, the busiest freeway in Southern California, did not in itself disturb me. The man was an excellent driver, very smooth. No problem either with the interior of his taxi being plastered with 8 × 10 glossies of snarling lions covered with religious messages about love, death, and the Lion of Judah. I am originally from New York City, where crazy cabbies spice the day. What did give pause, however, was the flawless way in which, when his cell phone rang, Rasta Cabbie would become James Earl Jones saying, "West Side Transportation, may I help you?" After wrapping up his office business, it was back to Jah and the lions and the bowing and the prayers.

I was headed to Guatemala, to meet with Mayan shamans who would explain the prophecies of 2012. When I mentioned this to Elia, my housekeeper, who is from El Salvador, she shouted, "No te vayas! Gangas! Think of your children. What if you don't come back?" and ran out of the room. Maybe Rasta Cabbie's prayer dance was some sort of tripped-out empathic blessing for a safe trip. Praise . . . Jah.

We pulled into LAX and on impulse I asked Rasta Cabbie if he'd ever heard about 2012.

"Educate me," he replied, as he hoisted my luggage out of the trunk.

"Well, people say big things are going to happen in 2012. Maybe, you know, the End."

"They always sayin' that. I was waitin' for that to happen in year 2000," he said, shaking his head sadly. But it was tip time, and Rasta Cabbie wanted to end on a positive note. "We keep workin' on things, and your year could be the one."

1

WHY 2012, EXACTLY?

Two hours' tromp through the tarantula/crocodile jungle where a recent *Survivor* series was set, past an ancient Mayan ball court where both losers and winners were sacrificed (that certainly would have boosted *Survivor*'s ratings) and then a steamy clamber up the hundred steep and crumbling steps of the 1,800-year-old ruin known as the Great Pyramid, the centerpiece of Mundo Perdido (Lost World), the oldest section of the Tikal ruins, was rewarded with the following: "The problem has got to be with your server. Call tech support and tell them to reconfigure . . . ," explained one twenty-something to the other.

Rip out their beating hearts, toss their lifeless carcasses down the stone steps, and chalk it all up as a human sacrifice to Bill Gates. Deep in the Guatemalan jungle, atop an ancient sacred temple, and these geeks still couldn't get their minds out of their computers.

I had gone to Tikal, where some of the most ancient Mayan prophecies originated, to get a feel for what, up until then, was just a mass of factoids—for example, that in the Mayan calendar the current age, known as the Fourth Age, began on August 13, 3114 BCE, which in the Mayan calendar is represented as 0.0.0.0.1 (Day One) and will end on December 21, 2012 CE, or 13.0.0.0.0 (Day Last). I could repeat that fact and many others accurately enough but, like twelfth-grade calculus (the derivative of n cubed is $3n$ squared, but what is a derivative, exactly?), I didn't really understand what I was saying.

The problem was calendars, to me a blah staple of contemporary existence. Navigating life without them would of course be unthinkable, but that's not going to happen, so why think about it? Apparently there once was a dispute between popes about how many days February and August should have, but that's all been settled for half a millennium. And at the stroke of midnight beginning 2006, the official atomic clock-keeper somewhere added a second for the first time since 1999 because the Earth's rotation is being slowed by the moon's increasing gravitational pull, which might be an interesting development if we had enough time in our busy lives to figure out why.

Fundamentalists insist that it's all in whatever their holy book might happen to be, but my visit to Mayan Guatemala was the first time I've ever been told that it's all not in their book but in their calendar, which is all I would ever need. The Maya love their calendars, see them as visual depictions of the passage of time, which is how life unfolds. They charted this unfolding with not one but twenty calendars, only fifteen of which have been released to the modern world; the remaining five are still kept secret by Mayan elders. Mayan calendars are pegged to the movements of the Sun, the Moon, and the visible planets, to harvest and insect cycles, and range in length from 260 days to 5,200 years and beyond.

In the Cholqij, the 260-day calendar that represents a woman's pregnancy cycle, and also the number of days that the planet Venus rises in the morning each year, each day is represented by one of 20 symbols representing spiritual guides or deities, called Ajau. The number 20 is sacred to the Mayans because a person has 20 digits—10 fingers to reach to the sky and 10 toes to grasp the ground. They regard the number 10, so significant to our mathematics, as half a loaf at best.

According to Gerardo Kanek Barrios and Mercedes Barrios Longfellow in *The Maya Cholqij: Gateway to Aligning with the Energies of the Earth*, 2005, thirteen forces influence the 20 Ajau deities. The number 13 is derived from the fact that there are 13 major joints (1 neck, 2 shoulders, 2 elbows, 2 wrists, 2 hips, 2 knees and 2 ankles), which serve as nodal points of bodily and cosmic energy. Thirteen forces times 20 deities equals 260 uniquely specified days.

The Mayan prophecies for 2012 are the province of the Long Count calendar, also known as Winaq May Kin, which covers approximately 5,200 solar years, a period the Maya call a Sun. In the curious Mayan reckoning, a year has 360 days; the remaining 5.25 days (4 x .25 accounting for the leap day) are considered "out of time" and are traditionally devoted to thanksgiving for the previous year and celebration of the year to come. Thus 5,200 of these Mayan years translate to approximately 5,125 of our Gregorian years. Since human civilization arose, we have passed fully through three Suns, and now are completing the fourth Sun, which will end on 12/21/12.

The Mayan counting system is primarily vigesimal, meaning that it relies on powers of 20, rather than 10. In this system the first placeholder (the one farthest to the right) is reserved for units of one day; the second for units of 20 days; the third for units of 360 days, or one Mayan solar year; the fourth for units of 7,200 days, or twenty Mayan solar years; and the fifth for units of 144,000 days, or 400 Mayan solar years. Interestingly, the number 144,000 figures prominently in Revelation, though it refers to the number of people who will be saved and serve the Lord during the Tribulation, the period of tumult that precedes the Second Coming of Christ.

In 13.0.0.0.0, the Mayan way of expressing the 12/21/12 date, the number 13 refers to the number of *baktuns*, periods of 400 Mayan solar years/144,000-day periods. The number 13, as noted, is sacred in their cosmology. One Sun works out to be 13 times 144,000 days, or 1,872,000 days long, 5,200 of the 360-day Mayan solar years. On the day after a Sun is completed, the Long Count calendar starts all over. Thus, December 22, 2012, the day after apocalypse, if such a day does come, will once again be the Mayan date, 0.0.0.0.1.

TIME'S ARROWS AND CYCLES

How did these people become so time-obsessed, out in the jungles and the highlands? It's not like the ancient Maya were catching planes or texting messages or even traveling anywhere.

"At first glance it might seem an exaggeration to attach so much importance to the sacred [Mayan] calendar. Yet anyone familiar with its role in the life of pre-Columbian Mesoamerica realizes that bound up with the calendar are many if not all of the more sophisticated aspects of the region's early intellectual life: the awareness of a cyclicity in the movement of celestial bodies, the evolution of mathematical skills by which they could manipulate the numbers derived from those cycles, and the development of a system of hieroglyphics for recording the results . . . with it must have come most of the trappings of civilization—astronomy, mathematics, writing, urban planning," writes Vincent H. Malmstrom of Dartmouth College.

We all know intuitively that time occurs in both lines, as though arrows were being shot, and cycles. Time's arrow refers to the simple fact that each minute follows the next in a straight line to infinity, or until Time ends altogether. Time's cycle refers to eternal continuums, such as day and night, winter, spring, summer, and fall, the waxing and waning of the Moon. Time's cycles and arrows can also be seen as reflecting different attitudes toward history: "those who ignore it are doomed to repeat it" (cycle) versus "yesterday's news" (arrow). I'd always tended toward the latter camp, that history, though it made for good stories, was past. But after separating from my wife at roughly the same age, and with more or less the same height, weight, and features as my father did when he was separated from my mother, the "doomed to repeat it" scenario did ring a bell.

Cultures tend to have predilections for either arrow or cycle. Contemporary postindustrial Western society certainly emphasizes the arrowlike onrush of time, passing faster and faster, blinking and beeping on watches, microwave ovens, cell phones, and turnstiles. An arrow-affinity speaks to a society's orientation toward change and progress, though sometimes to the point of ignoring recurrent, eternal values. This imbalance may well have resulted from our shift away from an agriculturally based economy, which of course is finely attuned to seasonal cycles, and toward industrial and informational production, which are less dependent on such natural rhythms.

The Maya were and are a cycle society. They see cycles in everything, and they love what they see. Progress is not nearly as important in their cosmic ethos as the serenity that comes from being in harmony with the eternal movements of Nature. The downside of course is that, being fixated on eternal cycles, the Maya might not notice the day-to-day changes occurring around them, a disregard that helps explain why, as many historians have noted, classic Mayan society degenerated and collapsed abruptly, without their ever having taken heed of the warning signs. Theories range from voluntary disengagement, meaning that the Mayans simply abandoned their cities and much of their lifestyle for (occult) reasons of their own, to internecine strife, to claims that the civilization never really fell so much as went underground.

The current scholarly bet is that environmental degradation did them in. Indeed, Jared Diamond's recent book, *Collapse: How Societies Choose to Fail or Succeed,* depicts the ancient Maya as the case study of what societies ought not to do to the local environment. Diamond methodically presses the argument that the Mayans overfarmed, deforested, and overpopulated their land. A 2004 NASA study confirms Diamond's condemnation. Pollen trapped in sediments taken from the area right around Tikal, dating back approximately 1,200 years, just before the Mayan civilization's collapse, indicates that trees had almost completely disappeared, replaced by weeds.

Diamond believes that the population density of the Classic Mayan civilization reached 1,500 persons per square mile. That's double the current density, for example, of Rwanda and Burundi, two of the most crowded and troubled nations in Africa. Warfare over scarce resources inevitably broke out, leading to a complete societal collapse—a peak population of between 5 million and 14 million in 800 CE tumbled 80 or 90 percent in less than a century.

"We have to wonder why the kings and nobles failed to recognize and solve these seemingly obvious problems undermining their society. Their attention was evidently focused on their short-term concerns of enriching themselves, waging wars, erecting monuments, competing with each other, and extracting enough food from the peasants to support these activities. Like most leaders throughout human history, the Maya kings and nobles did not heed long-term problems, insofar as they perceived them," writes Diamond.

The Mayan fall in power, prosperity, and population is quite possibly the most drastic any civilization has ever experienced. Does this invalidate their wisdom? It certainly doesn't recommend it, except possibly in the area of catastrophe, which historically they know better than just about anyone else.

SPINNING LIKE A TOP

Righteous indignation was still pumping my brain when it dawned on me that the exchange between those two computer nerds on top of the Tikal pyramid probably wasn't far off in spirit from the conversations that took place there originally. That very pyramid, in fact, was built specifically for astronomers to chart the heavens and keep track of celestial time.

Imagine two ancient Mayan astronomers, an elder and a younger, arguing about the stars on the eve of the vernal equinox. The elder observes that Polaris, the pole star of the Northern Hemisphere, is not in the same position it was on the vernal equinox thirty-six years ago, when he first started his observations. Over that time, Polaris has shifted in a westward direction, the elder declares, about the same distance as the width of the full Moon (roughly half a degree).

The younger astronomer recoils from the heresy. From time immemorial, an article of celestial faith is that, on any given day and date, the stars are supposed to be in exactly the same position from one year to the next. To say otherwise would mean that the great heavenly clock is not keeping perfect time.

Eventually the truth won out, and the elder's discovery was incorporated into the Mayan cosmology. Perhaps as long as two and a half millennia ago, their ancient astronomers sussed out the astonishing fact that slowly, inexorably, the heavens crank westward at the rate of about 1 degree every 72 years, and complete a full circle every 26,000 Mayan solar years, a period equal to five Suns. The next five Suns would see the polestar change from Polaris, also known as the North Star, to Vega, and then back again.

As we've been reminded over and over again since Copernicus, it's not the heavens but the Earth that moves. In fact the Earth spins like a top on its axis. Watch a top spin, and you will note that its axis slowly describes its own tiny circle. That process is called precession and is entirely analogous to what we perceive as the rotation of the heavens in the sky.

Precession seems to have been discovered more or less simultaneously by a variety of different cultures. Traditionally, credit for first understanding that the heavens are in fact a giant clock that takes eons to move around goes to Hipparchus, an ancient Greek astronomer who lived in the second century BCE. However, it now seems likely that the ancient Egyptians, Babylonians, and Sumerians had earlier grasped the concept.

Persian and Indian astronomers also knew of precession, perhaps via the ancient Greeks, and were so impressed with the fact that the heavens move ever so slowly in an incredibly huge circle that they attributed it all to a deity, Mithra. During the sixth century BCE, Mithraism spread rapidly throughout India, the Middle East, and Europe. At its peak in the second century CE, Mithraism was more widely embraced than Christianity throughout the Roman Empire. Its central doctrine sprang from the sacrifice of a sacred bull, from whose body all goodness sprang. Although Mithraism virtually vanished in the third century CE, with Islam eventually taking over in Persia later on, the Persian New Year is still celebrated on the vernal equinox, usually March 20, a festive holdover from Mithraic days.

Long-term cycles in the Earth's orbit and spin have more than cosmetic importance, according to Milutin Milankovitch, the brilliant Serbian astronomer. He examined three cycles, now known as the Milankovitch cycles, for their potential impacts on climate and catastrophe on Earth. The first cycle, known as eccentricity, simply accounts for the fact that the shape of the Earth's orbit around the Sun changes from being almost perfectly circular to slightly more elliptical, over a cycle that lasts from 90,000 to 100,000 years. Right now we are at the most circular stage in that cycle, meaning that there's only about a 3 percent variation in distance, and a 6 percent variation in received solar energy, between perihelion, the point where our planet is closest to the Sun, and aphelion, the point where our planet is farthest from the Sun. However, as the Earth's eccentricity cycle proceeds toward the point at which our orbit is most elliptical, the amount of solar radiation our planet receives at perihelion will be 20 to 30 percent greater than at aphelion. This will make for sharper seasonal contrasts and profound climate change. Milankovitch and his followers believe that previous ice ages are largely attributable to the Earth's eccentricity cycle.

Currently, perihelion occurs during the second week of January, shortly after the Northern Hemisphere's winter solstice. This works out nicely, at least for those of us in the northern half of the world, because we are getting that extra 6 percent boost of solar energy right in the dead of winter. This cozy situation won't last forever, Milankovitch observed. As the north polestar shifts from Polaris to Vega, the orientation of the Earth toward the Sun also changes, to a situation where perihelion will come during the Northern Hemisphere's summer solstice, meaning that we'll be getting our energy boost right in the dead of summer. And by then, 13,000 years from now, that energy boost will be two or three times as powerful as the boost we get today, because the Earth's orbit will have become more elliptical, making for greater differences between the amounts of solar radiation received at different points of the year. All in all, the Northern Hemisphere's summers will be hotter, and the winters, colder, making Southern Hemisphere real estate a good long-term buy.

WE ALL WERE TAUGHT that the Earth tilts on its axis, although just why the Earth's axis tilts at all rather than going straight up and down is still open to conjecture. Some believe that eons ago the Earth was bonked by an asteroid or another planet, knocking us cockeyed; others argue that the pull of the Sun's gravitational field, which would be strongest at the Earth's equator, where most of our planet's mass is, causes the Earth to tilt, "stomachward," to the Sun.

The tilt of the Earth's axis is what causes the seasons, since at different times of the year different parts of the planet lean either into or away from the sunlight. When the Northern Hemisphere is receiving direct sunlight, it is summer here, and days are longer than nights. At that time the Southern Hemisphere is receiving indirect sunlight, its winter, when nights are longer than days. On two days every year, the spring and fall equinoxes, all parts of the Earth have equal day and night.

In a cycle known as obliquity, Milankovitch discovered that, over the course of about 41,000 years, the tilt in the Earth's axis changes from 22.1 degrees to 24.5 degrees. Currently the Earth's tilt is 23.5 degrees. The greater the tilt, the more exaggerated the contrast between seasons. Imagine yourself on a cold winter night, standing over a campfire. Now lean your face closer to the fire. It gets hotter and your butt gets thrust farther out in the cold. This is just what happens to the Earth as its axial tilt becomes more pronounced.

Although some contemporary scientists quail at the notion, a preponderance of evidence from archaeological texts and artifacts clearly indicates that the ancients had a rudimentary grasp of astronomical cycles such as precession, eccentricity, and obliquity. This knowledge gave astronomer-priests an exalted position in their societies, for they were felt to be in communication with the gods. Knowing, for example, when Venus would rise was impressive not just as a calculation but more as a transmission of information from gods to priests and then to their followers. Thus the ancient Mayan revelations concerning 2012 were considered to be of divine origin.

For millennia, the night sky was humanity's readiest source of news and entertainment. The ancients watched the stars and the planets as avidly as we do television. Heavenly bodies were just that, bodies of the deities. Movements and changes in them indicated divine events. Ancient astronomer-priests took the art and science of sky-watching to the point where they could in fact predict the future—for example, lunar and solar eclipses. This required not only observation but also the mathematics necessary to correlate the movements of the Moon and Sun. Their sophistication gives lie to the condescending Hollywood gimmick wherein the white man, knowing that an eclipse is due, pretends to make the Sun disappear, thereby scaring the ignorant natives. The white men didn't know half as much as the ancients and indigenous peoples did about the stars.

Van Gogh looked up into the starry sky and saw the swirls of God's imagination. Three millennia earlier, Pythagoras listened to "the music of the spheres," silent to the ears but not to the immortal soul. You know those rare and wonderful moments when you're familiar with a composer but not the piece being played at the moment, and yet somehow you can sense where it's going and how it will end? Vivaldi's *The Four Seasons* and Bach's six *Brandenburg Concertos* are like that—listen to the first few, and the rest, though in no way derivative or redundant, just might unfold in your mind before the notes are played. Over the course of two dozen centuries of rapt connoisseurship, Mayan astronomer-priests developed an ear for how the music of the spheres played out, including the chords for disaster.

Prior to the 15th century, the Elders knew through the prophecies of the approaching invasion of the Spanish, which began on the first day of a cycle called the Belejeb Bolum Tiku (the Nine Darkness). This was a 468-year period consisting of nine smaller cycles of 52 years each, which lasted from August 17, 1519, until August 16, 1987 [the day of the Harmonic Convergence]. Because the Guardians of the Prophecies knew well in

advance of the approaching invasion, they had ample time to prepare their communities. They informed the people about the effect the invasion would have on them, the sacred land and their
traditional way of life. Part of the preparation included steps to
ensure the protection of all the records, including the codices [sa-
cred texts].

Most of the original Mayan codices, thousands of them, were burned
during the first weeks of the Spanish conquest in 1519, by order of the Roman Catholic Church. Father Diego de Landa, who supervised the burning,
was subsequently ordered by the king and queen of Spain to return to
Guatemala and write a book summarizing Mayan beliefs. The resulting text,
Relación de las cosas de Yucatán (Yucatán Before and After the Conquest), was
full of cultural and factual distortions, not the least of which was the opening declaration that all Maya revered Jesus Christ, of whom few, at that point,
had actually heard. Nonetheless, this book was the first text about the Maya
in any Western language and therefore became the basis for virtually all
Western scholarship on Mayan customs and beliefs, mistakes that have been
compounded ever since.

It is widely written that only four Mayan codices survived the Spanish
book-burning. What that means is that only four such codices are today
known to be in Anglo-European hands. Many more sacred texts were saved
by record-keepers and elders from different tribes who hid out in the mountains and remote areas. For more than twenty years, Gerardo Barrios visited
villages in Guatemala, El Salvador, Honduras, and Mexico, searching out the
descendants of these elders, some of whom still lived in the same caves in
which their ancestors escaped the conquistadors. As they write in *The Maya
Cholqij,* except for minor variations in language, "all of the calendars in use
by traditional Mayan communities match up and continue the accurate
record (count) of days that the Maya have been keeping for many thousands
of years." These texts were saved because the stars warned the Maya of the
impending disaster headed for their culture. Now the Mayan calendar tells
us that's what's ahead for the whole world.

On 12/21/12 our Solar System, with the Sun at its center, will, as the
Maya have for millennia maintained, eclipse the view from Earth of the center of the Milky Way. This happens only once every 26,000 years. Ancient

Mayan astronomers considered this center spot to be the Milky Way's womb, a belief now supported by voluminous evidence that that's where the galaxy's stars are created. Astronomers now suspect that there is a black hole right at the center sucking up the matter, energy, and time that will serve as raw materials for the creation of future stars.

In other words, whatever energy typically streams to Earth from the centerpoint of the Milky Way will indeed be disrupted on 12/21/12, at 11:11 PM Universal Time, for the first time in 26,000 years. All because of a little wobble in the Earth's rotation.

But why should a brief disruption of so distant a source as the center of the galaxy have any real consequences for our planet or its people? After all, we can go for days, weeks even, with no sunlight or moonlight without significant distress. The best analogy is the way that even a momentary disruption of electrical power can cause the clocks on VCRs and microwaves to go from keeping perfect time to blinking on and off meaninglessly until they are reset by hand. Our being even briefly cut off from the emanations from the center of the galaxy will, the Maya believe, throw out of kilter vital mechanisms of our bodies and of the Earth.

As I sidled gingerly down the steps of the Great Pyramid, I felt a pang for the chattering computer geeks. A sense of foreboding is in the air. We can all feel it, even those guys, and we can all find ways to deny those feelings, like by chattering nervously about anything but. Now it turns out that an ancient, obscure culture has for a good two millennia been predicting the date of apocalypse as 2012. There's internal logic and precision to the Mayan thinking, and they're sticking by the date. Denial has just become a little bit harder. Maybe a lot harder.

2

THE SERPENT
AND THE JAGUAR

"Count to 100 and ask me if I'm Peter Pan."

I'd pulled that old grade school prank too many times, struggling to keep a straight face while the sucker labored up through the count. Then finally the ridiculous question. The answer of course is no. Having spent thousands of dollars and hours to travel to Guatemala to climb up crumbling temples and now to meet with Mayan shamans, I wondered if it was my turn to be the sucker.

"Is the world going to end on December 21, 2012?"

"No. Not necessarily. It could all go quite smoothly, in theory," replied Carlos Barrios, debonair graybeard shaman of the Mam, one of twenty-six Mayan tribes in Guatemala. We were in Arbol de Vita, a Guatemala City vegetarian restaurant owned by Tony Bono, brother of the late singer and congressman Sonny Bono. The décor of this beautiful place can best be described as Zen/Maya; on the far wall, a contemporary abstract sculpture of a snake-bird keeps tempting my eyes away from the conversation. The figure is Kukulcán, the Mayan version of Quetzlcoatl, the supreme Mesoamerican deity of light and heaven.

"People today are terrified. We live in an age of nuclear weapons, terror, plagues, natural disasters. The year 2012 has become a magnet for all those fears. It is being taken out of context by those who wish to play on people's anxiety. We don't see it as a time of destruction but rather as the birth of a new system," explained Carlos in fluent Spanglish.

Birth, I observed, is accompanied by blood and agony.

"I have assisted at some births," the shaman, a professional healer, gently reminded me.

Carlos has been on the shaman path since he was seventeen. He was driving his father's car in the rural highlands when, through the dust, he saw several men in outlandish costume, Tibetan lamas, it turned out, performing a ceremony in the middle of a field. He jumped out of the car, ran over, and asked what was going on. A local shaman guiding the Tibetans tried to shoo Carlos away, but the lamas took pity. Carlos looked on in awe as they took four phallic-shaped ingots called *lingams*—one bronze, one copper, one silver, and one gold, about five pounds each—and buried them in the field.

The local shaman feared that Carlos would come back and steal the lingams, but the Tibetans were not worried.

"And, though I can't explain how it happened, my memory was somehow erased. In my mind I can see every detail of that ceremony, but since that day I have never been able to remember where is the field where those things are buried," Carlos averred.

He hung out with the Tibetan group for the next few days and learned that the lamas had been traveling a 10,000-mile path, repeating this ceremony at key geomagnetic nodal points along the way, in order to shift the Earth's sacred energy field from the Old World, Mount Kailas, also known as Kang Rinpoche or Precious Snow Jewel, in the Himalayas, to Lake Titicaca in Bolivia.

"My imagination was so excited by all of this that I decided then and there that I had to go to Tibet. But the visa and the ticket were very expensive, $10,000, so I sent to my father asking him for the money. He sent back a telegram that said, 'Ha ha ha!' " recalled Carlos, still chuckling forty years later.

Carlos is an Ajq'ij, a Mayan priest. He is trained in the use of earth, air, water, and fire, which, as in many indigenous traditions, are the four basic elements. Mayan shamans specialize in the use of one of these elements to heal, to predict the future, and to harmonize space. Carlos's specialty is fire, which re-creates the power of the Sun. Once again Kukulcán, the feathered god, snakes my attention as Carlos explains that fire is the door to other dimensions, the "stargate," or portal, through which the great wise men and women of the past are returning. According to contemporary Mayan belief, the ancestors already have started to come back and are mixing in with the population. They are not, Carlos says, interested in being recognized. By 2012 they will all have returned to fulfill the sacred mission of that momentous year.

"The Resurrection is being accomplished in the children being born today. Everyone who was ever born and died in the past will have returned by 2012," said Carlos, nodding when I asked if the world's population explosion is evidence of this mass reincarnation. He explained that there's always a reason a soul's incarnation cycle hits a dead end. For some souls, the stumbling block may be love, for others, courage.

"Between now and 2012, we all will have an opportunity to face, and overcome, the challenges to our personal evolution. Those who pass their tests will move on to enjoy an enlightened new era." Carlos added that those who fail will be stuck in this dimension for many thousands of years, at which point presumably they will get another chance to take the test.

Carlos's brother, Gerardo Barrios, coauthor of *The Maya Cholqij*, arrived. The sensitive caveman, long black hair and beard, not a hint of gray even though he is at least sixty, ordered a papaya soy shake. Carlos and I got another beer.

"Why are you writing a book about 2012?" demanded Gerardo, catching me off guard. Journalists, like all other prosecutors, are there to ask the questions, not answer them.

"It was the only thing I could do to feel better," I blurted. Nonsense vomited all over the lunch table. I covered by claiming that what I really meant to say was that I had been going through a divorce and found the work a welcome distraction. "I feel bad, so the rest of the world must die," I deadpanned.

Carlos smiled at the joke, but Gerardo wasn't so sure about the crazy gringo.

"The year 2012 is a seam in time, the juncture of two different ages," said Gerardo. "Death, possibly a great deal of it, will be part of that transition."

Is 2012 the equivalent of a "punctuated equilibrium," Stephen Jay Gould's description of the jumps and rough transitions by which evolution proceeds? Or, in the language of cybernetics, will we in 2012 make the leap from one steady state to the next? Gerardo nodded, but then modified his assent.

"The change will be gradual, more like the deepening of twilight than the flick of a switch."

Twilight fades fast, which means we'd better bone up for this test.

"The elders say that in the new time coming, after 2012, pain and happiness will be shared more and more. Mass communication makes us more like brothers and sisters, more like a family. In 2012 there will also be collective tests in harmony and understanding," said Gerardo.

I noticed that Kukulcán, the snake-bird, tugged at my eyes only when Carlos spoke; he seemed an other-worldly feathered serpent himself. By contrast, there is a darkness to Gerardo that is immensely appealing. He's the man you'd want along with you on a descent into the underworld, the realm of Balam, the black jaguar god in Mayan mythology.

Gerardo was, in fact, trained in darkness, placed in a tiny, pitch-black room deep underground for about two weeks. After a while he lost all track of time and space, night and day. He began to hallucinate and soon was able to visualize separately and distinctly hundreds of Mayan hieroglyphs used in their various calendars. In the black room he also heard a secret language he did not understand, though he was sure, if he paid attention, the language would one day help guide his prognostications.

Immersion in darkness is a theme in Mayan shaman training. Gerardo explains that sometimes the elders know a child is destined to become a great shaman while still in the womb. When the baby is born, they wrap thirteen bandages around its head, covering the eyes. Those bandages will stay on until the child is either nine or thirteen years old, loosened periodically as the head grows. The elders do this to sharpen the young shaman's other senses and also to enable him or her to read auras. In the final year of this imposed blindness, one bandage is removed each lunar month, so the eyes gradually get used to the light. The final bandage is removed inside a sacred cave, gently illuminated by candles. The first thing the young shaman focuses on is a Mayan codex, an ancient sacred book made of bark paper and deerskin and filled with colorful, intricate hieroglyphs, the same ones Gerardo visualized.

Lore has it that some ancient astronomers knew the sky so well that they could be kept in the dark for weeks until they lost all track of time and space. Then, on the first night they were brought out to observe, these astronomers could look up at the sky, sift through their memory, and tell the exact day and date by the position of the stars.

IT HAS ONLY BEEN for about half a century or so, a tiny fraction of human history, that mass communications have made it possible for us to respond emotionally to situations such as the Indian Ocean tsunami. Gerardo therefore observed that humanity is still in its infancy in its ability to empathize with the feelings of people far away. Nonetheless such empathy is crucial for the survival and transcendence of the species, which is why this skill is part of the coming reckoning.

The coming reckoning . . . are we talking Judgment Day?

Gerardo explained that in different stages of human history different messiahs come. This is an age where there will be lots of small guides, rather than one great messiah, according to the elders.

Gerardo booted up his awesome HP laptop, and the giant screen filled with images of elders, most men, mostly old, all with penetrating gazes. He and Carlos spent twenty years going from village to village throughout the Mayan territories in Guatemala, Mexico, El Salvador, and Honduras, searching out these elders. Some were still living in the same caves to which their ancestors had retreated to escape the sixteenth-century Spanish conquest that nearly extinguished the Mayan culture.

Gerardo has personally seen six codices saved from the Inquisition and is aware of the existence of several others. But the Mayan elders safeguarding these sacred texts have shown little interest in sharing the contents with Anglo-European scholars. Once burned, twice shy.

"It is not yet time to reveal their secrets," he huffed.

CULTURAL IMPERIALISTS

For a moment during lunch with Carlos and Gerardo at Arbol de Vita, I caught myself wondering if maybe the whole 2012 apocalypse thing weren't some sort of sneaky Mayan revenge hoax on the North. Lord knows they have their reasons. Sitting in the same restaurant where Sonny and Cher had actually once dined, it struck me that damn near every winner of every Academy Award, Emmy, Golden Globe, Grammy, People's Choice, you name it, had had their children raised, their homes kept, and/or their gardens tended for a comparative pittance by labor, legal and illegal, provided by Mexicans and Central Americans of Mayan or other indigenous descent, none of whom had ever received so much as one of those thank-yous that gush like cheap champagne throughout the award show proceedings. It's macabre, the dichotomy of how famous are Hollywood glitterati and how invisible are these people who hold the stars' lives (frequently very messy) together.

The Barrios brothers shrugged at Hollywood's condescension but boiled when I brought up the subject of archaeologists, who are a pet peeve with the Maya and many other indigenous cultures. Gross inaccuracies, cultural biases, self-aggrandizing personal agendas—the litany of complaints against archaeologists is endless, though in truth these are more criticisms of bad archaeology than of the discipline itself. For example, the image that emerges from centuries of "scholarship" on the ancient Mayan ball game, where two teams kicked a latex rubber ball up and down the field and attempted to put it through a hoop, is that the game was bloodthirsty, because it ended with the execution of certain players. In truth it was rather civilized. Instead of going to war over key trade routes, feuding parties would field their best teams. Losers would be sacrificed, preventing a much larger bloodbath on the battlefield. True, there were times when slaves were forced to play, and the losers killed for no other reason than blood sport, but that was an abuse of an otherwise reasonable war substitute.

There were also times when the winners met their death. For major celebrations, such as the end of a sacred fifty-two-year cycle, it was not unusual for Maya to volunteer themselves for sacrifice. What a way to go! In Tikal, for example, throngs of extravagantly clad citizens would fill the plaza, sitting before the steps of the pyramid of the Giant Jaguar, where priests dressed as animal and mythic entities performed rituals that taught basic Mayan precepts of cosmology and morality. The possibility of being sacrificed as part of this festivity drew more hopefuls than could be accommodated, so the candidates were divided up into teams that played the ball game. The winners got their reward.

What rankles deeper is the archaeologists' presumption that they are rediscovering "lost" cultures. How insulted would the average Italian person be if it were generally assumed that the fall of the Roman Empire meant that all of its linguistic, cultural, and technological accomplishments were lost by descendants too ignorant or careless to preserve the legacy? The Maya do a slow burn when self-impressed scholars elbow their way past native elders filled with the wisdom of the ages to foist their own interpretations on ruins and hieroglyphs.

The cultural imperialists' need to discover something that they are sure all indigenous sages have somehow overlooked can be vexing. For example,

John Major Jenkins, author of *Maya Cosmogenesis 2012*, a dogged freelancer who, by sheer dint of will, has thrust himself into the debate over Mayan history and culture, believes that Izapa, a little-known ruin just across the Mexican border, was the center of an empire that eventually gave rise to the Maya. Jenkins deploys page after page of complex and frequently far-fetched calculations using maps, calendars, and sky charts to bolster Izapa's prehistoric legacy. The Barrios brothers politely acknowledge the scholarly interest but strain with ennui when told by outsiders that Izapa is their true Vatican.

Archaeologists are impertinent. They compare cultures, and rate them on scales: technological development, legal codes, governance structures, and health and sanitary systems. Under the Mayan column, there's no check in the box marked "invented the wheel," a very touchy subject. Although the Mayan ancients grasped the concept of circles, cycles, and orbits more thoroughly than any of their contemporaries and in some ways better than we do today, they never translated that concept into actual, tangible wheels. Neither did any arches grace ancient Mayan architecture, roughly covering the two-millennia span from 100 BCE to 1000 CE, millennia after other cultures had discovered the beauty and utility of the curve.

All too often, archaeologists become lightning rods for a culture's insecurities. The Barrios brothers' feeble rejoinders that wheels wouldn't have worked so well in the jungle are easily dismissed when one visits the massive Mayan temples and wonders whether those diminutive slaves whose job it was to hoist 110-pound stone slabs might not have appreciated a few wheeled carts and a ramp. Then again, treatment of slaves is not a debate we Americans really want to get into.

Carlos shook his head wearily when I asked him about the band of Mayanists who believe that the big date is not 2012 but 2011. Led by Carl Johann Calleman, a cancer researcher associated with the World Health Organization who for years has dedicated himself to Mayan scholarship, these folks believe that the Mayans have miscalculated their own calendar. That's Calleman's hook, his scholarly identity. The mix-up is reminiscent of the dispute over calculating Y2K, and whether or not the millennium would turn on January 1, 2000 or 2001. Carlos, who is fond of Calleman, patiently explained that the Fourth Sun (Age) will end 12/21/12, on the winter solstice, which, as it happens, is expected to occur at 11:11 UTC (Universal time, formerly known as Greenwich mean time).

The First Sun, according to Carlos, began approximately 20,000 years ago, was dominated by female energy, and related to the fire element. The Second Sun was characterized by male energy and related to the earth element. The Third Sun was characterized by female energy and related to the air element. The Fourth Sun that we are just now completing has been dominated by male energy and related to the water element. On 12/21/12 we will enter the Fifth Sun, in which the energy is balanced between female and male. Related to the ether element, the Fifth Sun brings with it a subtler wisdom.

Fire, earth, air, and water are all known elements and together constitute pretty much the totality of physical life. But what is ether, exactly? Air you can't breathe? Thoughts? Even though I don't exactly understand it, to me the prospect that ether is the thematic element of the new age we are entering seems only good news. Unlike, say, fire, which lends itself to holocausts, or water, which can bring ice or floods, ether seems, well, ethereal— hardly the stuff of which apocalypse is made. However, it is the impending transition into such nothingness that causes consternation.

THE FEATHERED SERPENT and the Black Jaguar, which is how I thought of Carlos and Gerardo, devoted the prime of their life to revitalizing the Mayan network, to assisting elders in need, and to recovering codices and other artifacts. By heritage, training, and sheer dedication, they stand as the preeminent authorities currently writing and speaking to the outside world on Mayan culture, science, and prophecies. But as Gerardo ran down the elders' who's who on his laptop, I gather that, though he and Carlos have comparatively high-profile, well-remunerated positions interfacing with outsiders and the press, within the Mayan spiritual hierarchy they are midlevel at best. Quite in contrast, for example, to the Roman Catholic Church, where salary and benefits rise steadily from priest to pope, a Mayan shaman's spiritual stature has little to do with his or her material standing. Their kingdom, as Jesus might observe, is not of this world.

The 2012 prophecies, however, are very much of this world, space, and time. Despite the Barrios brothers' sugar-coating and caviling, done as much out of a fear of igniting a panic as a survival tactic to reassure themselves and their loved ones, the coming of 2012 does indeed portend catastrophe and dislocation on a global scale. The more time I spent with Carlos, the more open he became about his fears for that year. What really scared him was when an elder whom he especially reveres declined, during a sacred annual ceremony, to give his usual talk. The silence meant that there is nothing more to be said about 2012 and the dangers it holds.

It wasn't really until my final forty minutes in Guatemala that Gerardo opened up. It was 5:30 AM at the Guatemala City International Airport, and we were sitting cross-legged, me very stiffly and in pressed white pants, on the very dirty floor beneath a utility staircase. Gerardo had graciously come by to give me a farewell astrological reading. He handed me a soft little bag and instructed me to blow into it four times, once for each of the four directions and for the four elements of earth, air, fire, and water. A soldier/security guard with an automatic rifle slung over his shoulder took a sudden interest in our proceedings; I think he wanted to make sure I was exhaling, not inhaling. Gerardo paid him no heed, took back the bag, and tossed out its contents—red beans, jaguar teeth, various crystals—onto the multicolored cotton mat on the floor between us, and then studied them for a moment. Subject: my divorce. Response: philosophical.

My parents had separated when I was eight, and for the next two years I tried shuttle diplomacy to get them back together, but then my father died in a car crash. His car skidded on ice into an oncoming truck. I guess he was driving too fast because he was late to sell a man asphalt to cover his driveway. So I have tended to get death and divorce mixed up. Now, facing my own divorce, I couldn't seem to stop my internal life from orienting toward death. Didn't want to stop, in fact. Rather fond of the idea. Except you're not allowed to feel that way when you've got two young kids. Gerardo had picked up on all this and with his red beans and crystals and jaguar teeth somehow showed me the calm, not just resignation but true peace, that comes with accepting that death—of oneself, a loved one, one's marriage, the world—is not in one's control.

"Are we headed for divorce? From Time, from Nature, from our civilized lives? Is that what the 2012 prophecies are about?" I asked suddenly, catching him off guard with my question. Like the elders he revered, Gerardo declined to speak.

Black jaguars are the only cats that swim under water. They can stay down for quite a long time, but sooner or later they come up for air.

EARTH

I once was jilted for Mr. Spock, the supremely logical Star Trek character from the planet Vulcan, by Barbara Wetzel, the prettiest math whiz Junior High School 51 in Park Slope, Brooklyn, had ever seen. Barbara, who lived with her aunt and her sister in an apartment above an embalming parlor, and who had the faint, sweet scent of formaldehyde in her long blonde hair, informed me that she had lost all interest in human men.

I continued to watch Star Trek, though now in the way that my father used to follow the New York Yankees, rooting for them to lose. The best hope of seeing Mr. Spock, Captain Kirk, and all their cohorts fry was if the starship Enterprise came under ferocious attack when its protective shields were down. This did happen from time to time, with Klingons, Romulans, and all manner of other angry aliens pounding away, though invariably by the end of the hour crew and ship escaped intact. The difference, of course, is that spaceships can fly. We landlubbers have no choice but to stick around and take our beating.

3

THE MAW OF 2012

The great white shark's breath was so bad I could smell it under water. Or maybe it was the divers on either side of me in the cage, puking their guts out. The scarred-up shark, a two-ton thirteen-footer with huge double rows of blood-stained steak-knife teeth, had the crazed look of a predator that hadn't evolved during his 400 million years terrorizing the oceans. It bonked its hideous snout against the dented-up cage, then chomped down on the bars. The image of Ulysses, lashed to the mast, listening to the sirens' insanely beautiful song, did not flash through my mind. But there was that kind of thrill, a moment stolen from the gods.

Good practice, I mused, for gazing into the maw of 2012.

I was off the southern coast of South Africa and was scheduled the next day to visit the Hermanus Magnetic Observatory, where geophysicists are examining the California-sized cracks that have been opening up in the Earth's protective magnetic field. Next stop was Johannesburg, to meet with 123Alert, a group of psychics who have an impressive record of forecasting earthquakes, volcanoes, and the like. My little adventure on *Shark Lady* was just for larks, until I looked inside those great white jaws.

For the first time in the year or so since I had been researching the horrors of 2012, I stopped and gave thanks, in this case for the bars on the cage. My life had been so safe and healthy that I had been taking it entirely for granted that I would live to see 2012, when I would be fifty-eight, which happens to equal the average male's lifespan in the former Soviet Union, meaning that about as many men won't live that long as will. Back home in Beverly Hills, where everyone is young and lives forever, talk of death is tantamount to hate speech. But in South Africa, about one in five adults, skewed toward the young, will likely die by 2012 anyway, even if the year passes without an apocalyptic peep.

"Thank you, India. Thank you, Providence."

Back safely aboard the *Shark Lady*, my mental boom box played Alanis Morissette's edgy elegy to thankfulness. Her song challenges one to think of new things to be thankful for . . . metal cages, for one. The great white shark's indignation at having a gourmet dinner, me, dangled unattainably in front of its nose, reminded me of one of my favorite questions: If God or some other Higher Power in which you believed, offered to give you exactly what you deserve, no more, no less, for the rest of your life, would you take the deal?

That question goes straight to a person's self-concept. Most people say they'd take the deal, some agreeing so emphatically that they think it's a trick question. Of course! Wouldn't anybody? Folks who pounce on this offer tend to believe that their life, or life in general, is a raw deal. They would welcome "justice" with open arms. Others get pedantic and argue that by definition we all are getting exactly what we deserve, because we deserve exactly what we're getting. God is just, so however we are treated must therefore be just— that sort of circular reasoning.

Personally I'd turn down the deal in a heartbeat. I know I've got it good. Maybe better than I should.

What if this proposition were offered to humanity as a whole? What if the heavens opened up and God/Higher Power/Mighty Space Alien offered our civilization exactly what it collectively deserved? No more, no less. How would you vote? Does humanity deserve all the stress and heartache? All the violence, disease, and degradation?

"*24 Children Are Killed in Baghdad Car Blast: U.S. Soldiers Are Said to Be Giving Out Candy, Toys.*" According to the report from the New York Times

News Service, the soldiers, one killed and three injured, had entered the Baghdad neighborhood to warn residents that there was an explosive device in the area.

On the other hand, do we, the same species that sets off those car bombs, deserve all the wonders of romance, the beauty of Nature, the sweet love of little children? Couldn't really say who deserves what. But I'd bet heavily that, in a straight-up global vote, the Almighty's proposition to give humanity exactly what it deserved, no more, no less, would pass overwhelmingly. Why wouldn't the teeming Third World billions give a thumb's up to economic justice? Let's see, people in the West make, say, ten to a hundred times as much money, live 50 percent longer, get to travel, educate their kids, and even get to obsess, as comedian Chris Rock reminds us, about things like lactose intolerance.

Something else new to be thankful for—that I find the prospect of Apocalypse 2012, that bloody-toothed doom crashing its snout against the fragile cage of human existence, bone-terrifying. In Capetown, where I had stayed several nights in a converted prison, I couldn't help but wonder who might in fact welcome the final cataclysm? No one in his right mind, of course. But then, lots of people aren't in their right mind, some through no fault of their own. It was easy, sitting in that cell of a hotel room, to imagine a political prisoner so angry and afraid that he or she would welcome destruction, as long as it included the jailers, would welcome perhaps a moment of freedom as the prison walls tumbled down and the very ground beneath them split apart.

As the great white swam off in search of baby dolphins and tasty seals, I recalled having read that there's a small concentration of iron magnetite in the shark's brain that enables it to navigate the Earth's magnetic field. If those cracks get any bigger or the magnetic poles flip, that shark will never find me or any other prey. It won't know where it's going.

The Earth's magnetic field—yet another thing for sharks, and humans, to be thankful for.

SHIELDS DOWN

Hermanus is a picturesque town on South Africa's southwestern cape. With sheltered coves, this breeding ground for southern right whales is considered the best land-based whale-watching spot in the world.

I had spotted at least ten when Brian, a local bay activist, came over and asked me if I'd like to see the whales closer up. I nodded, and he produced a convoluted instrument, put it to his lips, and blew his own special version of the sound that Gabriel will one day make to signal Judgment Day. After several tortured, bellowing, oddly musical blasts, a half dozen or so whales swam toward us; one, a seventy-tonner by Brian's reckoning, spouted its hello. He kindly sold me the horn, which he had made out of dried kelp, for forty rand (seven dollars). He then recommended the *Shark Lady* excursion to me.

"Keep all digits inside the cage," he advised with a solemn nod.

Brian loves his whales, would rather Jonah's fate than seeing harm come to them, so I refrained from asking what he thought about the fact that the magnetic field that guides these great whales in their ocean travels from Antarctica to Hermanus and back was weakening. Sooner or later it will dwindle beyond their capacity to sense it. That question was reserved for Pieter Kotze, one of the geophysicists I had come to Hermanus to see.

Kotze is a calm character, a man who fully expects to live a good, long, quiet life. When I visited him at the Magnetic Observatory, a lovely greenspace on a hill overlooking the bay, the geophysicist hospitably gave me a tour of the quaint laboratories full of state-of-the-art computers analyzing data transmitted from electromagnetic probes buried deep underground. The Earth's magnetic field originates from the spinning of its molten iron core, which is why the sensors are buried. Kotze asked me if I had any children and what their birthdates were. He then excused himself and after a moment popped back with two seismograph-style readouts of how the Earth's magnetic field had behaved on the day each child was born.

Kotze's work is as disturbing as his manner is gentle. He has meticulously chronicled the recent depletion of the Earth's protective magnetic field. After the tour, he patiently brought me up to speed on what it all means.

We can't repeal the law of gravity, a good thing, since without also repealing the law of inertia we'd all go flying off the Earth. Neither can we repeal the laws that govern electricity or magnetism. But there's no law that says the Earth has to have a protective magnetic field shielding us from excessive proton and electron radiation from the Sun that would spur an epidemic of cancers in human beings and many other species, disrupting the global food chain. The glut of solar radiation would also block out cosmic rays, highly energetic particles and waves from outer space that scientists now believe account for much of the cloud formation around the Earth. Clouds, particularly low-lying ones, block out infrared radiation—heat—from the Sun and help keep the Earth's surface cool.

The Earth's magnetic field deflects solar radiation and channels it into belts that harmlessly circle our planet's outer atmosphere. None of our neighboring planets has such a field, at least not nearly to the extent that Earth has currently. In fact our strong, well-functioning magnetic field is not to be taken for granted, particularly because it appears to be in the process of reversing and perhaps diminishing to the point where it will offer little or no defense from the Sun's depredations.

Traditional geology has it that the Earth's magnetic field, or magnetosphere, is generated by the spinning of the planet's core, a mixture of molten and solid iron that essentially acts as a Moon-sized dynamo, creating a giant electromagnetic field that squirts out of the poles, coalesces in the same basic pattern that iron filings do around a bar magnet, and bulges far into the atmosphere. Kotze explained that the interplanetary magnetic field (IMF), essentially the magnetic field that emanates from the Sun, also influences the magnetosphere's size and shape. Sometimes the IMF energizes the magnetosphere with inputs of solar energy. At other times the IMF presses upon the Earth's magnetic field, condensing, distorting, and even tearing holes in it.

The anthropocentric point of view of the magnetosphere's function is that its primary purpose is to prevent potentially lethal incoming solar radiation from reaching the surface of the Earth. There is no scientific reason that our planet should be taking precautions to defend its living organisms. There are, however, valid religious reasons why God might protect His creation in this manner. If nothing else, chalk it up to damn good luck that the Earth has a molten core that for the past 5 billion years has spun a powerfully protective magnetic field thousands of times stronger than those of any of the other inner planets—Mercury, Venus, or Mars. Without that shield, life on Earth would probably never have had a chance to evolve.

The Earth's magnetosphere channels incoming solar radiation into two belts, known as the Van Allen radiation belts, discovered in 1958 by Explorer I and Explorer II upper-atmosphere research missions, under the direction of the now-legendary James A. Van Allen. The Van Allen belts are wide, ranging in altitude from 10,000 to 65,000 kilometers (6,000 to 40,000 miles), and are densest at about 15,000 kilometers (9,000 miles). The inner belt is composed mostly of protons, and the outer belt is composed mostly of electrons. When these belts reach capacity, radiation spills out, strikes the upper atmosphere, and fluoresces, causing polar auroras. Because the Van Allen radiation belts pose certain hazards for astronauts passing through them, as well as for satellites, several far-fetched proposals have been made for draining them off. The good news is that that will not be necessary if the magnetosphere responsible for channeling charged particles to them ceases to function. The bad news, of course, is that it won't be just the astronauts who are worried about lethal radiation if we mess with those belts.

The Core, a Hollywood feature film that Kotze enjoyed but dismissed scientifically, portrays the catastrophe that would happen if the Earth's core stopped spinning as a dynamo and therefore stopped creating the planet's electromagnetic field. Of course, for the core to stop spinning, the Earth would have to stop spinning on its axis, which would have even direr consequences, such as wreaking havoc on staples of our existence like the seasons and even day and night. But the film nonetheless introduced the valid notion that our magnetic shield is vital to our existence, and that it just might be getting a bit threadbare.

Scientists are basically clueless as to why the magnetic field is dwindling. Speculation ranges from turbulence in the interplanetary magnetic field to chaotic fluctuations in the fluid dynamics of the Earth's molten core. It could be haphazard, or strictly cyclical. Kotze confirms, however, that it has all happened before.

There is much jittery speculation about whether or not the dwindling of our planet's magnetic field means that the poles are about to flip. Compasses that now point north would point south, and vice versa. The first step in a magnetic pole reversal is the weakening of the overall field, such as we are now experiencing. Imagine one sumo wrestler on top of another, pinning him down. Before the bottom wrestler can switch positions and end up on top, there has to be a lot of twisting, wrenching, grasping. For several moments at least, the wrestlers will be side by side, before the reversal is complete. Same idea with the magnetic poles switching their positions, except that instead of moments, this reversal process will take hundreds of years, during which time the Earth will have multiple magnetic poles, and compasses will point north, south, east, west, and all points in between. Birds will get lost; sharks like my frustrated great white will swim aimlessly; frogs, turtles, and salmon will be unable to return to breeding grounds; and polar auroras will flash at the equator. In all likelihood, the weather will get even weirder, with the tangle of magnetic meridians playing hob with the direction and intensity of hurricanes, tornadoes, and other electrical storms.

Studies of ice core samples and sediments extracted from the ocean floor indicate that the magnetic poles last reversed themselves about 780,000 years ago. At that historical depth in the geological record, magnetic rocks and bits that would now face north faced south, and vice versa. For the next thousand or so years, magnetic specimens were found facing in all different directions, before aligning themselves in the north-south pattern that may finally be eroding today.

> As to the changes physical again: The Earth will be broken up in the western portion of America. The greater portion of Japan must go into the sea. The upper portion of Europe will be changed in the twinkling of an eye. Land will appear off the east coast of America. There will be upheavals in the Arctic and in the Antarctic that will make for eruptions of volcanoes in the torrid areas, *and there will be shifting then of the poles— so that where there has been those of a frigid or the semitropical will become the more tropical, and moss and fern will grow.* (italics mine)
>
> EDGAR CAYCE, *Reading 3976–15, January 19, 1934*

In this reading, given while he was in his "sleep state," Cayce is said to have channeled the Archangel Halaliel, an enemy of Satan and a companion of Christ. To be sure, his most dramatic predictions have not yet come to pass and, may it please the Lord, never will. However, two important elements, the shifting of the magnetic poles and the Earth growing warmer, are indeed occurring. How, one wonders, could Cayce, lying on a bed in a New York apartment in 1934, know what the best and brightest of our scientists, with their state-of-the-art technology, are just now coming to terms with?

Perhaps it's just the law of averages. Predict enough different kinds of catastrophe, and odds are that some of them will come true. But in the Hutton Commentaries, an unusually scholarly Web site, geologist William Hutton argues that even small shifts in the location of the magnetic poles can have significant consequences. Hutton points out that there are two basic types of pole shift possible: "In the first mechanism, all the layers of the Earth remain together and the axis and the entire spinning globe tilts relative to the plane of the Earth's orbit around the Sun," writes Hutton. He explains that this type of shift results in the north and south poles moving relative only to the position of fixed stars. This would not result in any seismic or volcanic disturbances, since the Earth's crust, mantle, and core are not moving relative to each other. Unfortunately this is not the type of pole shift we are experiencing, argues Hutton, because the only movement of the poles relative to the Earth in this scenario would be due to the infinitely slow, millimeter-by-millimeter creep of continental drift.

By contrast, the poles appear to be moving far more rapidly, skittering across northern Canada and Antarctica by 20 or 30 kilometers per year, respectively. Hutton believes that we are experiencing what is known as a mantle-slip mechanism, which refers to the slipping of the Earth's mantle and crust over the liquid core, or over some malleable surface just above the core. This process could easily cause the "wandering pole" syndrome observed with some alarm over the past decade.

"This type of mantle-slip pole shift also causes the pre-shift equator to move over the surface of the Earth," writes Hutton. "As the pre-shift equator moves into new regions of Earth's surface, these regions begin to experience changes in centrifugal forces and sea levels. This leads to new distributions of land and sea, and to crustal tectonic movements." Such movements, Hutton contends, could presage the kind of seismic and volcanic calamities that Cayce predicted.

Kotze, the South African geophysicist, isn't so sure that a pole reversal is imminent. Neither is Jeremy Bloxham, of Harvard University, who believes that the process may take a millennium or more. Bloxham nonetheless warns that the weakening of the magnetic field, even well short of a complete pole shift, will diminish its shielding effect. We will be much more susceptible to the radiation constantly bombarding our planet from space, much the way that in *Star Trek* the starship *Enterprise* was at its most vulnerable when its shields—energy fields that protected the ship—were down. The *Enterprise* and her crew always managed to escape immolation, disintegration, and all other consequences of the death rays shot at them, because that's the way television series go. Of course the Earth and her inhabitants come with no such guarantee of a happy ending.

HOLES

The European Space Agency will send out Swarm, a trio of research satellites that will thoroughly examine the Earth's magnetic field, from 2009 to 2015. Well before then, scientists better unravel why the field has been cracking for as much as nine hours at a time. The largest, a 100,000-mile crack known as the South Atlantic anomaly, opens up over the ocean between Brazil and South Africa. The danger, quite simply, is that this hole, which may well be the first of many, is a gaping chink in our armor against solar and cosmic radiation. A number of satellites passing through the South Atlantic anomaly have already been damaged by solar outbursts penetrating the diminished magnetic field, including, ironically, a Danish satellite designed to measure the Earth's magnetic field.

"The more advanced the community is, the more vulnerable it is to the effects of outer space," declared Kotze in our interview. Kotze is most worried about the vast networks of power grids that keep the world electrified. They are very susceptible to solar outbursts, particularly those now regularly penetrating the South Atlantic anomaly. Blackouts are always inconvenient, and in nations such as South Africa, where there is a high crime rate, they are a threat to social order.

The South Atlantic anomaly is unsettlingly close, just a few degrees north, to the infamous hole in the stratospheric ozone layer over Antarctica. It could well be that the two holes are related. The dwindling of the Earth's magnetic field may in fact be causing the ozone layer to dwindle as well. Kotze explains that when proton radiation from the Sun penetrates the Earth's magnetic shield, the chemistry of the atmosphere is affected: temperatures spike and stratospheric ozone levels plummet.

A brief history of the ozone controversy is helpful here. In the mid 1970s, James Lovelock, a maverick English atmospheric chemist, took his prized invention, the electron capture detector, a palm-sized radioactive ionization chamber capable of sniffing out ionized gases at the parts-per-trillion level, and sailed from Britain to Antarctica and back, analyzing the air along the way. At every point, even thousands of miles out into the open ocean, chlorofluorocarbons (CFCs) were found to be present—gases that are exclusively man-made. Apparently CFCs never decompose. Lovelock published his results in *Nature*, though with no thought as to what the impact of these peculiar aerosols might be.

Later that year, Ralph Cicerone and his colleague Richard Stolarski of NCAR, the National Center for Atmospheric Research in Boulder, Colorado, drew the scientific world's attention to how chlorine catalyzes the destruction of ozone, showing how one slippery and promiscuous chloride ion can slide in and out of hundreds of thousands of unstable ozone molecules, lingering just long enough to shred their bonds. In 1974 F. Sherwood (Sherry) Rowland and Mario Molina of the University of California, Irvine, demonstrated that CFCs, as carriers of chlorine to the stratosphere, were therefore a grievous threat to the stratospheric ozone layer. Rowland and Molina delineated the complex reaction sequence of the CFC destruction mechanism, and for their work they shared, along with coresearcher Paul Crutzen of Germany's Max Planck Institute, the 1995 Nobel Prize in Chemistry.

Depleting the ozone layer makes the atmosphere more permeable to the Sun's ultraviolet (UV) rays. It is important here to note that the increase in UV radiation reaching the Earth's surface is almost completely a function of the thinning of the atmosphere's defenses, a thinning caused by man-made gases. One shudders to think of the impact that surging solar UV rays pouring through the Earth's cracking magnetic field might have on our planet, particularly as we head toward the unprecedented turmoil of the solar maximum projected for 2012.

As most sunbathers have come to know by now, ultraviolet radiation can be broken out into two basic categories: soft ultraviolet (UVA), which does not burn the skin, and hard ultraviolet (UVB), which does. Increasing exposure to UVB radiation has elevated the incidence of skin disorders ranging from sunburn to melanoma and also of certain eye disorders. The health risks are considerable (to fair-skinned people, anyway), but what really hit home in our Sun-worshipping culture was that the Sun was no longer to be revered but feared. It was the end of an era begun in 1920, when Coco Chanel admired the bronzed sailors on the Duke of Westminster's yacht and then "invented" the fashionable tan by getting one herself. That era climaxed with an impish puppy pulling down the swimsuit bottom of the brown-as-a-berry Coppertone girl, a.k.a. Jodie Foster, exposing her bright white butt.

Now little white butts will burn faster than ever, because more and more cosmic rays are slipping through the Earth's magnetic shield, shredding ozone molecules in much the same way chlorine atoms do, by splitting the bonds between ozone's oxygen atoms. Of course, prospective CFC manufacturers might seize upon this finding as an opportunity, arguing that the impact of the dwindling of the Earth's magnetic field is what has been depleting stratospheric ozone. According to this line of thinking, CFCs may be less harmful than previously thought and therefore needn't be regulated so stringently. Environmentalists will counter that we should control what we can, in this case CFCs, to control damage to the ozone layer.

Clearly there is an adverse synergy developing between the weakening of the magnetic field and the depletion of the ozone layer, resulting in greater threats to human and environmental health. Yet few if any forums regularly bring together scientists specializing in the ozone hole to meet with scientists specializing in the Earth's magnetic field.

Carlos Barrios had his own take on the dwindling of the Earth's magnetic field. I asked the Mayan shaman if it didn't seem somehow suicidal, the Earth dropping its guard against hot lover Sun. Barrios looked at me with a jaded pity reserved for the interminably naïve.

"Have you ever had fungus on your skin?" he asked.

No, but I knew people who had. Lots of itching and bad-smelling lotions and ugly red blotches. "You mean we are the fungus, on the skin of the Earth?"

Carlos nodded that it was certainly possible, then added, "The treatment for skin fungus is to lie out in the Sun."

4

HELLFIRES BURNING

As we bobbed in a boat in a cove in the harbor of Heimaey, in southwestern Iceland, Hjalli, the captain, beseeched God's mercy for twenty minutes, begging that we might survive. We were about to set out for Surtsey, the youngest island in the world, named for the giant in Icelandic mythology who keeps the hellfires burning. When Surtsey emerged on November 14, 1963, the ocean boiled. The crew of a fishing boat in the area was too busy with their nets to notice anything, until a great black column rose out of the water, blotting out the horizon off the bow. Four more years' of volcanic eruptions formed the pudgy half-square-mile teardrop, which ever since has refused to submerge, despite being in one of the stormiest places on Earth, with more than 200 gale-force days annually, and waves up to 85 feet high.

From its birth, Surtsey was set aside as an ecological preserve, completely off limits to tourists, with no permanent structures of any kind, including docks or even moorings, permitted. It had taken me a year to get all the permissions to visit, and this after having been personally invited by Iceland's president, Vigdis Finnboggadottir. But when Hjalli finished praying and then blew his trumpet to let Gabriel know we might be seeing him soon, I wondered aloud whether we shouldn't just putt-putt around the harbor for a while and agree among ourselves that the Surtsey trip had been, well, beyond words.

My assignment was to write a magazine piece about how Surtsey developed an ecosystem, how a steaming hunk of lava rock surrounded by salt water comes alive. Birdshit, in a nutshell. I understood from the research that ocean-going birds eat fish and poop on the island, providing fertile spots for seeds blown in the wind and washed up from the ocean surface. Sea sandwort, a tenacious green succulent with white and yellow flowers, is usually the first colonizer, because its hammocklike structure traps the sand that the plant's roots need to keep from being blown away by the fierce ocean winds. All in all quite interesting, but not worth perishing at sea.

Snorri, the naturalist assigned as my guide, confirmed that Hjalli was the only captain experienced enough to land safely at Surtsey and advised me to overlook the fact his other boat had been smashed to fiberglass bits on the jagged shoreline. All hands, after all, had survived. Besides, added Snorri, a Christian fundamentalist, the next life is bound to be better than this one.

That 1993 trip to Surtsey turned out to be an unforgettable afternoon, sliding down black *aa* lava tongues as long as the escalators in the London Underground, climbing into huge vagina-shaped craters and rolling around in their luscious green moss, warming our hands over fissures sputtering sulfurous steam.

After ducking a raven with a wingspan as wide as my arms, Snorri asked me if I was hungry, then had himself a good laugh. The joke went right over my head, but Rasta Cabbie would have gotten it right away. Jah, I now realize, refers to Elijah, the Old Testament prophet whose story in I Kings is what Snorri was referring to. Under the wicked King Ahab, the Hebrews had turned away from God, who decided to punish the land of Israel with three and a half years of bitter drought. God instructed Elijah to tell Ahab of the drought that would come, then flee Ahab's wrath by hiding at the brook of Cerith, where Elijah was kept alive by ravens bringing him food.

It wasn't until I met the Barrios brothers in Guatemala a dozen years later, however, that I understood the true import of that trip: Surtsey was my first glance at what our future might look like physically: a postvolcanic desert. What I had always back-of-the-minded as a remote if cataclysmic possibility may well be impending, as Yellowstone, one of the world's largest supervolcanoes, gets ready to blow.

WHEN YELLOWSTONE BLOWS

How betrayed would you feel if Yellowstone, America's first, most famous, and most exciting national park, erupted and put an end to our society? Better to feel betrayed than to feel nothing but burning sulfur choking the life out of your lungs.

The question is not if Yellowstone is going to blow, or even when. It's not as though there's an alarm clock inside the world's most dangerous supervolcano, ticking toward some preset explode date. The fact is that it could erupt at any time, filling the atmosphere with sulfuric acid and ash and plunging the planet into a nuclear-winter-type catastrophe, savaging economy and agriculture so severely that civilization might never reemerge.

The supervolcano scenario is quite similar to the nuclear winter envisioned by Carl Sagan and the TTAPS group in the late 1970s. Each type of explosion creates concentric circles of doom. Ground zero of course would be scorched and purged of all living things. A Yellowstone eruption would cause much of Wyoming and Montana to quickly resemble Surtsey: black steaming rubble awaiting the droppings of birds.

The next largest circle of nuclear winter hell would be poisoned with radioactive fallout, as it might be from the Yellowstone supervolcano, which sits atop enormous uranium reserves. Winds would blow such fallout thousands of miles, fatally sickening human beings and livestock. Thyroid cancer would be the quickest illness to strike.

Those two circles of doom, infernal though they may be, would pale in lethality when compared to the effects of the ash cloud borne across the North American continent on westerly winds. Ash would clog up jet engines, render the air unbreathable, and in the long run blot out sunlight and cause temperatures to plunge and therefore crops to fail and economies to falter. The Northern Hemisphere, where some two-thirds of the world's land mass and population is located, would see its interdependent societies collapse, as food became scarce and darkness plunged frightened psyches into depression. With the world population at nearly 6.5 billion, who can say what end of carnage and warfare might result from this calamity?

Yellowstone has had at least 100 major eruptions, three of which were unfathomably massive, each large enough to cause hemispheric calamity were it to occur today. The first eruption occurred 2 million years ago, and it was followed by another one 1.3 million years ago. According to a March 2006 *Nature* cover story investigating puzzling magma flows in and out of the supervolcano, Yellowstone's most recent full-scale eruption occurred roughly 640,000 years ago and coughed out about 1,000 cubic kilometers (218 cubic miles) of ash into the atmosphere. This would be enough to bury the entire continental United States at least one meter deep in soot and cinders. It's as though the Great Lakes were all filled twice over with ash, which was then dumped out over the continent. This is easily enough ash to block out the sunlight for the better part of a decade.

Crude math yields a periodicity of 600,000 to 700,000 years for the supervolcano's eruptions, meaning that chronologically we are right on schedule for the next big blow.

More important than probability statistics is what's happening underground. A guest on a BBC Horizon documentary on the Yellowstone supervolcano, Professor Robert Christiansen of the U.S. Geological Survey, recalled that he had found many rocks made of compressed ash in his visits to Yellowstone, but for years he could never find any evidence of the volcano from which they must certainly have erupted. He consoled himself with the thought that it must be very tiny. That thought exploded in 1993 when NASA, testing some infrared photographic equipment designed for scanning the Moon, took heat signature photos of Yellowstone and revealed the largest single caldera ever discovered. Calderas are large underground depressions containing magma, a mixture of solid and liquefied rock and highly combustible volcanic gases. The Yellowstone caldera is unbelievably huge, the size of the city of Tokyo, some 40 to 50 kilometers long and 20 kilometers wide, the molten, beating heart of Yellowstone Park.

Subsequent geological surveys revealed that the caldera (which means "cauldron") has risen about 3/4 meter since 1922, filling with magma and getting ready to explode. Compared to other geological timescales, such as the millimeter-per-century continental drift and the virtually imperceptible weathering of mountains, such a change is downright tumultuous.

As Robert B. Smith, a geologist and geophysicist at the University of Utah, reports, this supervolcano's topographical distortion is so pronounced that Yellowstone Lake, which sits atop the caldera, is now actually tilting because of the bulge. Water is draining out at the south end, inundating trees that just a few years earlier grew normally out of the soil along the shore.

61

HELLFIRES BURNING

"It would be extremely devastating, on a scale we've probably never even thought about," says Smith of the coming Yellowstone eruption. Estimates of its explosive force range up to the equivalent of 1,000 Hiroshima-style atomic bombs—per second. This would be roughly the equivalent of all the violent energy ever expended in all the wars ever fought, per minute.

"I'm not sure what we would do," says Steve Sparks, a professor of geology at the University of Bristol, of a Yellowstone eruption, "except stay underground."

Supervolcanoes are quite different from the cone-shaped volcanoes with which we are familiar. They are depressions in the ground, from several hundred to more than a thousand miles deep, usually with complex networks of rivulets, tubes, and tributaries through which magma can flow. There is some disagreement on how supervolcanoes' deep structures operate; most appear to channel magma and explosivity from deep in the mantle, the thick, liquidy layer between the crust and the core that accounts for most of the Earth's volume.

Supervolcanoes are far more powerful than conventional volcanoes. By definition, they measure 8 on the volcanic explosivity index (VEI), which runs from 1 to 8. Like the Richter scale for earthquakes, VEI is logarithmic, meaning that each number indicates a blast ten times greater than the preceding number. Mount St. Helens, considered a large blast, was a VEI 5.

Other supervolcanoes around the world include Kikai Caldera in Ryukyu Islands, Japan; Long Valley Caldera, California; La Garita Caldera, Colorado; and Camp Flegrei, Campania, Italy. A supervolcano blast at Lake Taupo, New Zealand, in the year 186 CE, devastated New Zealand's northern island. Compared to a Yellowstone, however, the Lake Taupo eruption would be but a puff of steam.

To understand how supervolcanoes work, imagine a blazing abscess moving and growing under your skin, suffusing the flesh below with fiery pus. In geological terms, this abscess is known as a hot spot, and the Earth's skin, or crust, is moving over it. In *Windows into the Earth*, Robert Smith and his cowriter, Lee. J. Siegel, former science editor of the *Salt Lake Tribune*, explain that most hot spots are "columns or plumes of hot and molten rock that begin 1,800 miles underground at the boundary between Earth's core and lower mantle, then flow slowly upward [because heat rises] through the entire mantle and crust."

Hot spots are typically located at the boundaries of tectonic plates, which essentially float on seas of molten rock. Hot spots tend to be located along the seafloor, since most of the Earth is covered by water. The molten churn of these hot spots is composed largely of basalt, which tends to seep and flow rather than explode.

"Of the roughly thirty active hotspots on Earth, almost all except Yellowstone are beneath oceans or near coastlines or other boundaries between tectonic plates. The best known of the other hotspots are those that produced Iceland [including Surtsey], the Hawaiian Islands, and the Galapagos Islands," write Smith and Siegel.

The Yellowstone hot spot, by contrast, is smack dab in the middle of our continent. It's quite far from any ocean or plate boundary, the closest one being roughly at the Pacific coast. And the Yellowstone hot spot does not extend down into the Earth nearly as far as the others. According to current estimates, it reaches a depth of only about 125 miles, less than a tenth the normal. Thus its impetus does not come from the Earth's molten core. Rather it appears to have been formed largely from the heat produced by decay of vast amounts of uranium and other radioactive elements in the region, heat that then melts iron-rich basalt rock, great blobs of which periodically plume to the top.

"The molten blobs of basalt heat overlying crustal rock, creating a 'magma chamber' in which silica-rich crustal granite partially melts, formally a molten rock known as rhyolite when it erupts. . . . Because molten rhyolite is thick and viscous, major eruptions from the Yellowstone hot spot have been explosive, unlike the basalt that erupts more gently from oceanic hot spots [such as Surtsey]," explain Smith and Siegel.

Think of the way a pot of thick stew left on the flame too long might all of a sudden splatter the kitchen, where a pot of watery soup would bubble and boil over less explosively.

The Yellowstone hot spot appears to have formed some 16.5 million years ago, beneath the areas where Oregon, Nevada, and Idaho meet. Since then it has had several dozen eruptions, each of which would have devastated any civilization that might have existed at the time.

A macabre illustration of Yellowstone's handiwork was discovered by Professor Michael Voorhies, of the University of Nebraska. After heavy rains, Voorhies went to the little town of Orchard, Nebraska, to do some fossil hunting. What he found was an archaeologist's dream and everyone else's nightmare: hundreds of skeletons of rhinos, camels, horses, lizards, and turtles, most in their prime, all killed abruptly 10 million years ago, almost certainly coinciding with a Yellowstone blast. The skeletons in this mass catastrophe were covered with a white film, forensic evidence that the animals died of something akin to Marie's disease, a lung disorder most likely contracted by the inhalation of volcanic ash.

Slowly and steadily, the murderous hot spot has shifted some 500 miles in a northeasterly direction to its present location in northwestern Wyoming, where its caldera bulges menacingly under Yellowstone National Park. Like any other abscess continuously scratched and abraded, the hot spot will pop and spew. Once emptied, it will settle back down and then slowly refill over the next 600,000 years or so, until it explodes again. Also like any other abscess, there is not necessarily any preset, optimal explosion moment, just a range of "ripeness."

The swirl of data and innuendo surrounding Yellowstone's current seismic activity is almost as thick as the molten rhyolite that one day will pop its cork. Anecdotal reports of impromptu police action, unannounced trail closures, discoveries of heat and seismic sensors, and other appurtenances of heightened watchfulness abound on the Internet, contrasting starkly with official, what-me-worry? attitudes.

An uptick in Yellowstone's seismic activity would be a tipoff that an eruption was coming; dozens of seismographs have been installed in and around the park, to relay as early as possible the slightest bad news. Swarms of tiny earthquakes, a chemical change in the composition of lava, gassing from the ground, cracking of the land—all these are potential signs that eruption is imminent. A rapid and substantial rise in the elevation of the caldera, swelling, it would be assumed, with magma and volcanic gas, would be an obvious tipoff.

The problem, as the producers of the BBC specials on Yellowstone quickly found out, is that for reasons unspecified, much of this data is unavailable to the public. For example, numerous reports of a 100-foot bulge in the bottom of Lake Yellowstone, a normally cold, high mountain lake whose water temperatures have somehow hit the mid-80s, have gone unconfirmed, and also unchallenged, by park authorities. For the most part, however, one is forced to rely on unofficial information sources. According to Bennie

LeBeau, from the Eastern Shoshone Nation in Wyoming, a number of new steam vents have formed along the Norris Geyser Basin, where soil temperatures reached nearly 200 degrees in 2003, at which point the entire 200-square-mile basin was closed.

At issue here is the public's right to knowledge that could have bearing on their safety versus the government's duty to protect its citizens from the dangers of panic. But the fact of the matter is, the eruption could come virtually unannounced: "The only reasonable conclusion that one can come to in studying the current Yellowstone caldera environment is that there is no current way to reasonably and accurately forecast the eruption of the Yellowstone caldera," writes R. B. Trombley, a volcanologist with the Southwest Volcano Research Center in Arizona.

So what could make the Yellowstone caldera pop? To answer this question, it is necessary to understand the inner dynamics of its magma chamber, a banana-shaped structure whose uppermost tip is believed to be about ten kilometers beneath the surface of the Earth. Since subterranean forays into such molten, explosive media are impossible or ill-advised, even for robotic probes, the best information on Yellowstone's magma dynamics comes from historical data on similar explosions.

The most recent analogous supervolcano eruption occurred 3,500 years ago in Santorini, Greece. Though much smaller in scale than any Yellowstone eruption is anticipated to be, it can provide lessons. According to Steve Sparks, the Santorini eruption hurled 2-meter chunks of rock 7 kilometers or more, at supersonic speeds. Research revealed that inside Santorini's caldera was a great deal of liquid magma, in which highly volatile volcanic gases had dissolved. Sparks led a team that constructed a trillionth-scale model of the supervolcano's eruptive forces and found that, as the top of the chamber opened, as it would at the beginning of an eruption, sudden depressurization inside the caldera caused the volcanic gases dissolved in the magma to expand and explode violently, jetting the magma into the air.

Sparks has shown that a supervolcano caldera, filled with liquid (magma), does not act like a water balloon, which slowly oozes out its contents upon being breached. Rather, it acts like a gas balloon, which explodes with a pinprick. That finding must be considered bad news, because the most obvious, practical measure to prevent or delay Yellowstone's eruption—drilling a hole into the caldera and releasing some pressure—would have precisely the opposite effect. It would set off the eruption to end all eruptions.

Could a well-executed nuclear strike by a major power such as China or Russia, or even by a rogue nation such as North Korea or Iran, potentially pop the Yellowstone balloon? How about a terrorist attack, the likelier scenario? A message accompanying the March 11, 2004, Al Qaeda bombings in Madrid (which occurred exactly 911 days after 9/11) referred to "the black wind of death" that will blow over America. One chilling fantasy making the rounds is that the terrorist act alluded to in Al Qaeda's message, a project purportedly 90 percent complete, would entail the insertion of a thermonuclear device into the Yellowstone caldera, igniting the supervolcano, filling the atmosphere with ash, and shoving much of satanic North America into the dustbin of history.

As there appears to be no current way to defuse or dissipate the supervolcano's eruptive mechanism, the scale and the volatility being just too immense, we are forced to rely on responsive rather than preventive measures. An extensive early warning system has already been installed, with ultrasensitive seismic and thermal sensors placed in and around the park. Where all that data goes and, more important still, who makes decisions based on that data, remains unclear.

Local, state, and particularly federal government officials all have the Yellowstone situation firmly on their radar, both in terms of civil preparedness and patrolling and surveilling against potential foul play. Or at least that's what I would have assumed, until the BBC began looking into the matter in March 2000. In the measured, eminently factual style for which the British network is known, a group of scientists who have studied Yellowstone for years laid out their case, which was only politely received. So BBC executives decided to try again, and in March 2005 they presented a two-part, four-hour docudrama, later shown on U.S. television, dramatizing the impact of a Yellowstone eruption. Perhaps the most tragic outcome expected is the prolonged failure of the Asian monsoons, which would likely plunge the most populous region of the world into famine and disease.

The docudrama's producers and principal scientists subsequently presented an executive summary of their findings to the Federal Emergency Management Administration (FEMA), in Washington, D.C. FEMA acknowledged receipt, and a spokesperson admitted that not much had been done to prepare for such an eventuality. Several months later the southern United States was hit by the killer hurricanes of Katrina, Rita, and Wilma, and FEMA had more than it could possibly handle.

No doubt the BBC's Yellowstone report was duly filed, perhaps next to the much bigger dossier compiled for the Long Valley in central California, a caldera that the USGS (United States Geological Survey) describes as "restless" and "actively rising." Formed 760,000 years ago, when a supervolcano eruption blew out 150 cubic miles of magma, covering much of central California with ash that blew all the way to Nebraska, the Long Valley eruption, though smaller than Yellowstone's most recent, was still 2,000 times the size of Mount St. Helens. It could probably plunge the Northern Hemisphere into volcanic winter. In response to what the USGS with surprising candor refers to as Long Valley's "escalating geologic unrest," extensive new monitoring, assessment, and emergency procedures have been instituted.

The USGS reports that geologic unrest at Long Valley began in 1978 and then spiked two years later with swarms of earthquakes:

> The most intense of these swarms began in May 1980 and included four strong magnitude 6 shocks, three of which struck on the same day. Immediately following these shocks, scientists from the U.S. Geological Survey (USGS) began a reexamination of the Long Valley area and detected other evidence of unrest—a dome-like uplift in the caldera. Measurements made by these scientists showed that the center of the caldera had risen almost a foot since the summer of 1979 after decades of stability. This continuing swelling, which now totals nearly 2 feet and affects more than 100 square miles, is caused by new magma rising beneath the caldera.

Whatever the individual odds of Yellowstone, Long Valley, or Lake Toba erupting in our lifetime, that number must be multiplied by twenty, thirty, or more to reflect the number of supervolcanoes around the world that we know of. Each of these calderas is capable of wreaking havoc similar in scale to Yellowstone. Then multiply that figure by another ten or twenty to include the calderas that we do not know of, particularly those that lie under the oceans. Then yet another multiplier, unknown but most likely large enough to bring the risk of global cataclysm unacceptably high, to account for the fact that a growing number of scientists are coming to believe that the likelihood of volcanic eruptions is increased by global warming.

VOLCANO COOLERS

"There is evidence that there have been several episodes of increased volcanism worldwide in the past and a possible link with climate change. Whether climate change is a cause or effect of variations in the rate of volcanism remains an intriguing question," writes Hazel Rymer, of the Open University in the United Kingdom, in *Encyclopedia of Volcanoes*.

Intriguing indeed. If, as Rymer suggests, global warming may exacerbate volcanism, which has the net effect of cooling the planet with ash and aerosols that shield the Earth's surface from the Sun, then expect more and larger eruptions.

Volcanoes appear to constitute a global cooling mechanism, thermostatically moderating the periodic temperature spikes that occur over the eons, quite possibly including the global warming happening today. This is the basic perspective of the Gaia hypothesis, which holds, in a nutshell, that the Earth more closely resembles a living organism that adjusts and regulates itself than it does a rock on which life just happens to be a passenger, or a geological machine that runs on automatic. As I explored in my first book, *Gaia: The Growth of an Idea*, James Lovelock and Lynn Margulis, and now their many adherents, believe that our planet's climate regulates itself to maintain conditions conducive to the continuation of life. Not that the Earth in any way consciously "thinks" that because it is growing hot it must shoot off volcanoes. If this mechanism exists, it operates homeostatically, akin to the unconscious wisdom of the human body, which when it overheats begins to sweat, without any conscious thought to do so.

So, has volcanism increased as the climate has warmed?

This question, unfortunately, is beyond the pale of science for the simple, startling reason that we have no way of knowing for sure whether volcanic activity is increasing, decreasing, or remaining the same. In fact, we have no clear idea how many volcanoes there are in the world. Surface volcanoes total slightly more than 1,000: 550 active volcanoes (meaning they have erupted within historical times, roughly the last 3,000 years) and 500-

plus dormant volcanoes (meaning they have erupted in the period between the last Ice Age, 11,500 years ago, and the beginning of historical times). These totals do not include submarine volcanoes, which are believed to be far more numerous for the good reason that most of the Earth's surface is covered by water. No one has a handle on how many submarine volcanoes there are.

A cheap scare tactic, common on the Internet, equates the number of volcanic eruptions with the total number reported, which has skyrocketed, for the simple reason that the technological array of eruption-detecting devices, from satellites to seismographs placed around the globe, has expanded exponentially. But that doesn't necessarily mean there are any more volcanoes, just more volcano reports. The same goes for earthquakes. Not long ago there were only a few seismographs operating in the United States. Now there are more than twenty in Yellowstone National Park alone, not to mention the thousands in California. The number of earthquakes reported has increased commensurately, but that means nothing in terms of actual trends, as many prophets of doom would have us believe.

Surface volcanic eruptions, including everything from simple lava flows to mega-explosions, range in duration from days to millennia. For example, the volcano that created Surtsey spewed lava for four years. Stromboli, off Italy's Aeolian Sea shore, has been steadily erupting for the past 2,500 years. Thus, any attempt to equate the number of eruptions with the total amount of volcanic activity is a crude approximation at best.

Volcanologists therefore rely on a commonsense method of gauging activity. They reason that really big volcanoes would not have gone unreported, no matter the period in history. By piecing together historical reports of action and damage, they can estimate the magnitude of these major events. Therefore, comparing the annual number of volcanoes over VEI 4, trends can be discerned. Mount St. Helens was a VEI 5, which scale of eruption occurs every decade or so. By this big-volcano-counting measure, global volcanic activity has remained relatively steady for as long as estimates can discern. The problem with this measure is that it is too crude to reflect regional variations, because the statistical sample of major volcanic eruptions is too small. Three more VEI 4 eruptions than average over the course of a year in, say, Alaska would spike a graph sky high, but it wouldn't necessarily mean much.

So we are riding the horns of a dilemma, as my grade school teacher used to say. We believe that global warming may be forcing more volcanoes and also supervolcanoes, but we don't have the means to measure that trend one way or the other. At least not scientifically.

Anne Stander, a psychic who lives outside Johannesburg, South Africa, heads a group called 123Alert, which specializes in predicting seismic and volcanic activity. They have a commendable track record, meticulously documented, of predicting earthquake swarms, and they have been tracking, often in advance, a virtually unreported proliferation of microquakes along the southern California coast.

Stander's most famous prediction was the March 8, 2005, eruption of Mount St. Helens which, since it was preceded by no scientifically detectable seismic activity, took geologists by surprise. All the more embarrassing because the plume of ash and steam was 30,000 feet high.

"Helen makes a lot of noise, like a Jack Russell terrier. But her husband, Rainier, is the one to watch out for," says Stander.

Is Stander guilty of anthropomorphizing spiritless geological processes? Perhaps, but her character assessment of Rainier is shared by conventional scientists: "More recently, it has been realized that Mt. Rainier poses a significant hazard to the growing population of the greater Seattle area, and more alarmingly that mudflows from Mt. Rainier could devastate the southern Puget Sound *without* [italics theirs] an eruption and with little warning," write Tony Irving and Bill Steele, volcanologists at the University of Washington.

Stander senses a correspondence between the Yellowstone supervolcano and the Cascade Range volcanoes of Mount St. Helens and Mount Rainier, though she can't say exactly what that relationship is. She firmly believes, however, that any and all drilling in the Yellowstone region should cease immediately. Why tempt fate? Unfortunately the Bush administration is doing just that, having authorized the drilling of an additional 10,000 oil wells in Yellowstone, in addition to the 5,600 already there. Just as disconcerting are several recent scientific research proposals to drill up to a dozen 2-to-3-kilometer holes in some of the most seismically sensitive parts of Yellowstone, to test out a hypothesis that the supervolcano's hot spot is fueled by mantle plumes.

Overall Stander foresees a rise in seismic and volcanic activity, particularly along the western edge of the Pacific rim, from Alaska on down through

California and Mexico. The peak will come in 2011, a reminder that 2012, though clearly the target date, will not pop out of the void but rather will culminate a range of cataclysmic processes. "I have said before that we need to worry about 2011, because all the signs will be there to let us know what 2012 has in store for us. The number 2011 brings a bigger danger of pain than 2012," says Stander.

LIGHTING THE VOLCANO FUSE

Politically, volcanoes rank below earthquakes, and way below hurricanes, on the blameworthy scale. For example, the Bush administration took a great deal of heat for not properly anticipating the impact of Hurricane Katrina. Had it been an earthquake that devastated New Orleans, Bush would not have been criticized nearly so severely, because we accept that no one can see an earthquake coming. However, had San Francisco been leveled by a seismic event, FEMA would have been strictly judged on its preparedness, because earthquakes, though not predictable, occur frequently enough in certain regions that government agencies should be prepared to respond.

Volcanoes however are anybody's guess and therefore no politician's responsibility. Funds being short and time being scarce, "volcano preparedness" is one of the first issues to get buried.

Life is imperfect, and government much more so. What with all the pressing concerns of crime, health care, tax burdens, and such, why carp about volcanoes? Even with the frankly disturbing confluence of supervolcano rumblings and the tie-in with global warming, this is one potential catastrophe source I was prepared to pass up until I put it together that a period of global warming immediately preceded and quite possibly caused the cataclysmic supervolcano eruption of Lake Toba, Sumatra, some 74,000 years ago.

The Toba supervolcano is believed to have ejected some 6,000 cubic kilometers of lava, ash, and debris, filling the air with sulfuric acid, choking untold humans, animals, and plants. But that was only the start of the chaos.

The most recent Yellowstone-scale blast, Toba created a nuclear-winter-type cooling that plunged temperatures anywhere from 5 to 15 degrees C in less than a decade. Such a precipitous climate change today would devastate the global food chain. How, for example, would the Florida citrus crop fare if, in a matter of months, local temperatures fell to Vermont levels? The chain reaction would domino throughout the global ecosystem. Birds and fish dependent on plants and algae would suddenly find themselves bereft. Livestock grazing would be disrupted, as snows came much earlier and pushed farther south. Grain production, the backbone of the agricultural economy because corn, wheat, and rye are massively consumed directly and as feed, could be slashed to a fraction by the freeze.

The Toba eruption was what science writer Malcolm Gladwell might categorize as a tipping point, chilling the climate just enough to push an already cooling planet into an Ice Age. Ice core samples taken from Greenland suggest that the Toba eruption "was followed by at least six years of such volcanic winter conditions, which were in turn followed by a thousand-year cold 'snap,' " writes Bill McGuire, a University College of London volcanologist, in *A Guide to the End of the World*. This cold snap in turn became the Ice Age from which, McGuire contends, the Earth fully emerged only 10,000 years ago. It is within only the last fifty years or so that the Earth's surface temperatures have returned to pre-Toba levels.

McGuire confirms the emerging scientific consensus that, as a consequence of the Toba blast, the world's population decreased abruptly, perhaps by 90 percent or more, down to a miniscule total or 5,000 or 10,000 individuals, and remained at that level for up to twenty millennia. In other words, our species almost went extinct as a result of the Toba explosion. A proportionate die-off today would result in 4 to 5 billion deaths.

Funding research on the relationship between global warming and volcanic eruptions may seem to have all the urgency of a dust ruffle. But if recent rising temperatures have set a Yellowstone, Long Valley, Toba, or some other volcano or supervolcano ticking, we need to know about it as soon as possible. Imagine if, in the past, rising climate temperatures had somehow set off a nuclear warhead, or every nuclear warhead ever made. No time for business as usual, no time for budgeteers and researchers to do their usual dance, to study, assess, debate, cogitate, reallocate, experiment, publish, bicker, and then stick it all on a shelf somewhere. By then we'd be waiting for birdshit to start the life cycle all over again.

5

CROSSING ATITLÁN

Crossing Atitlán, a vast and gorgeous lake nestled among three central-casting volcanoes in the Guatemalan highlands, I kept leaning over the rail of the motorboat and scooping out water to see if there weren't some natural dye that made it so blue. Noting virtually no development on the lake that Aldous Huxley called the most beautiful in the world, on a lake that puts Tahoe to shame, the real estate calculator embedded in my California mindset clicked merrily along until Lord Byron, the voluble young shaman-in-training who was my guide and interpreter for the day, pointed to a shoreline stained by a receding waterline. A 7.5 earthquake that struck on February 4, 1976, killed 22,000 people in Guatemala, left a record 1 million homeless, and cracked open the bottom of Lake Atitlán. The water is slowly draining away.

We got off the boat at Santiago Atitlán, probably the largest indigenous town in Central America, 37,000 people, 95 percent Tz'utujil (CHOO-too-heel) Maya. It's a lively, disheveled place, full of beaming children bouncing along in open-air vehicles that never saw a car seat, women in multicolored *huipils* (native weavings done in a tradition that dates back 2,000 years) balancing bundles on their head. We were met by Juan Manuel Mendoza Mendoza, thirty-two, a rising star in the Mayan spiritual hierarchy. Manuel, a short, handsome, powerful-looking father of four, cleansed my soul several hours later when he spat a mouthful of very cheap rum into my face.

To be fair, we all got a little tipsy at the first religious service. Manuel, Lord Byron, and I had kicked off the afternoon by venerating Maximón, the playboy saint, prophet, priest, and magical protector of indigenous people, who also happened to be a raging drunk and womanizer. The Spaniards, as the story goes, would execute Maximón at noon, but he would always be back in the town square next morning, sometimes hung over. At Cofradia Apostol Maximón, a pagan/Catholic storefront shrine, I stuck a twenty-dollar bill on Maximón's statue and toasted him with a glass of beer. Out of Maximón's (wooden) mouth always sticks a lit cigarette, and one of the priests, whose main job was to flick off the ashes, seized a sacred moment and passed out the Winstons. I took my first drag since quitting at 9:15 AM on September 1, 1985, then choked back the giggles as a healthier-than-thou American couple, probably from Oregon, recoiled in horror when the hospitable priest offered them cigarettes as well.

Worshipping at the feet of a playboy womanizer was a story I might have to edit for my wife, I thought, and laughed out loud. Later I learned that it is perfectly acceptable, even expected, to laugh during a Maximón ceremony. Quite a refreshing contrast to worship based on the Bible, the Quran, and the Bhagavad Gita—there's not a single intentional laugh among them. After toasting a couple more times and smoking our butts down to the filter, Manuel, Lord Byron, and I made ready to leave, though not before a 100 percent polyester red-and-multicolored Happy Holidays scarf was tenderly pulled from the idol's neck and knotted like an ascot around mine.

I asked Manuel about 2012 and he focused on the positive: "2012 is very important because it is a time when the elders from the past will return to make a communication between the heart of humanity and the heart of the Earth. It's the beginning of a new era of peace, harmony, love, and union. But at the same time there exists the possibility for manipulation. Evil enters where there is space for it to squeeze in. In order to defend against evil, we have to do a lot of ceremonies to determine the right path."

Why is it so hard to believe that 2012 could possibly be the dawn of a glorious new age? Is it because rosy scenarios do not satisfy any deep psychological need? If most of the evidence pointed to 2012 being the start of some-thing great, rather than horrible, no profit-minded publisher would have taken on this book. Are we too jaded and cynical to believe in a coming utopia? Or are we so satisfied with life that any major transformation, no matter how it's billed, is threatening? Maybe all we really want is more of the same, plus a few sweet tweaks.

The prospect of Apocalypse 2012 ultimately serves as a projective test for anyone who contemplates it. In a "last shall be first and the first shall be last" kind of way, the ones most open to post-2012 reality are those who have the least to lose in the coming upheaval. Does it take a relative detachment from material possessions, as many of those living in Santiago Atitlán seem serenely to have achieved, to have faith that Mother Earth will always pro-vide, if not for us personally, then for humankind or for the wiser species that evolves from our seed? Is what appears to be the threat of 2012 actually a challenge to be great, to face cataclysm squarely in the eye, and in so doing to uplift ourselves to a higher level of being—braver, kinder, and closer to the Divine?

Out into the street and then up the steps of the Catedral de Santiago, where mass was under way. We peeked into a hole, called R'muxux Ruchiliew, "the Navel of the Earth," dug right in the center of the sixteenth-century church. It's a portal leading to the underworld realm where sacred ancestors live. Once a year, at the stroke of midnight that commences Good Friday, a wooden crucifix is lowered through the hole into the ground, thus "planting" Jesus and enabling his rebirth on Easter.

Behind the altar, Manuel showed us a series of wooden panels intricately carved with distinctly non-Christian symbols and images, including Maxi-món, who is also known as Mam, the spirit of death. The shaman explained that in traditional Mayan theology creation is not a one-time act of the past but an ongoing process that must be actively sustained. If the cycles are bro-ken, existence will cease. Humanity's role in the grand scheme of things is to perform the rituals and make the sacrifices necessary to make sure the Sun keeps crossing the sky and the seasons keep changing.

"When Maya need rain, we pray for the rain and, sooner or later, it comes. We make our rain," explains Manuel.

This is incorrect. Maya do not make their rain any more than Belgians do. The Sun would cross the sky, and the seasons would come and go, no matter what rituals were or were not performed. We know this from modern science, and it is not just a matter of our belief versus their belief. It is a matter of fact versus error. If this book were basically anthropological in nature, more space would be devoted to the examination of Mayan beliefs and rituals. But the purpose of this book is to assess the importance of the year 2012, particularly with regard to any dangers it might present to the reader.

And yet there is something so commendable in the indigenous spirit of ecological partnership that it feels reckless not to take another look. But can genuine wisdom reside within gross factual inaccuracy?

Consider the opening of Genesis, the story of how God created heaven and earth in seven days, a story that, for the purposes of this book, is considered grossly inaccurate. The basic Darwinist model of evolution proceeding through natural selection has been proven a thousand times over by more than a century and a half of rigorous scientific research and is respectfully accepted here. Reliable evidence for the creationist view, that God/Yahweh purposefully created everything in six days because it pleased Him to do so, is, to say the least, scant. Yet this in no way diminishes the immortal genius of Genesis. What a remarkably insightful projection, 3,000 years before Darwin, of how life on Earth arose. From the planet's fiery emergence in the darkness of space, to the accumulation of liquid water, to the rise of plants, animals, and human beings, the foresight of Genesis is positively supernatural, even if the facts and reasoning have since been corrected.

For twenty-eight of the past thirty centuries, Genesis proved more accurate than just about any competing scientific theory. A similar case can be made for the prescience of Mayan cosmology. I asked Manuel what he thought would happen if the Earth-worshipping rituals were to cease. He fell mute at the thought. My guide/interpreter jumped in: "Life without the rituals would be like driving a car without shock absorbers. The ride would be a lot bumpier, but you could still get where you want to go. Until you hit a hole in the road," declared Lord Byron. Manuel grinned, and agreed that Earth worship soothes the planet.

Will 2012, I wondered, be like a giant pothole in the path of Time?

"That's the possibility for which we must be prepared," affirmed Manuel.

On our way out of the church, we stopped at the plaque dedicated to Father Stanley "Francisco" Rother, a missionary priest from the Roman

Catholic Archdiocese of Oklahoma City, who led the Santiago Atitlán congregation for thirteen years, building schools and a clinic, halving the infant mortality rate, and enabling local craftsmen to restore the cathedral interior with indigenous (what some might call pagan) artifacts, such as the carving of the playboy saint, Maximón. Rother, apolitical, even mildly right-wing, was gunned down, and stabbed to death in his rectory on July 28, 1981, at the age of forty-six. The death squad was working for General Fernando Romeo Lucas Garcia, the Guatemalan dictator whose policy was to eliminate leaders of indigenous communities, regardless of their politics. Rother's body was shipped back to the United States, though not before his heart was removed and buried in the church.

The cathedral looks out over a dusty paving-stone plaza that serves as the town square. Late on the evening of December 1, 1990, several young women crossing the plaza were hassled by government soldiers hanging out there. The women protested, some villagers threw stones at the soldiers, who then pulled out their weapons and fired, killing one person. Local folks, outraged, began ringing the cathedral bells to assemble the town. Several thousand gathered, then sometime before dawn on December 2, chanting for the soldiers to go home, they marched to the military garrison. When they arrived at the gates, the soldiers opened fire, leaving eleven dead and forty or more wounded.

Within hours, press from the mainland arrived on the scene, and their photographs of men, women, and a child lying dead at the gates of the garrison were proof that even the authorities could not deny. To the astonishment of just about everyone, President Garcia agreed to permanently remove all military forces from Santiago Atitlán. The town declared that it was, in effect, withdrawing from the Guatemalan civil war that had by then been raging for twenty-four years. It was the first place in the nation to do so. Soon after the troops departed, the people made a Peace Park, with plaques left where each martyr fell and an eight-ton marble sculpture chiseled with the text of the president's letter promising to remove the army and investigate the incident. While constructing the park, workers uncovered a mass grave, the likely container of the up to 800 other villagers who had been "disappeared," but the president's office threatened to send the army back if the grave was excavated, so it was left, and it stands today as a big hole in the middle of the park.

Guatemala's thirty-year civil war ended in 1995 with 100,000 dead, another 100,000 disappeared, 1 million displaced, 440 villages erased. The war has had a number of unintended consequences, for example, a brisk trade in stolen children, sold for adoption to Americans—in many cases, Guatemalans believe, for immoral purposes. A generation of chaos left the country wide open to become a prime transshipment territory for drugs. Much as the Roman Empire paid its soldiers in salt, the cartels pay theirs in cocaine, which has become Guatemala's alternate currency. One gram of cocaine is worth seven dollars, according to current exchange rates.

Not surprisingly, Guatemalan voters have become rather cynical. They once elected the comedy team of Taco and Enchilada president and vice president of their country, with 70 percent of the vote. The comedians, who had formally withdrawn their names from the ballot two weeks before election day, and who thus won in a write-in campaign (particularly noteworthy in a nation where literacy is far from universal), declined to take office.

I felt obliged to take a picture of the town square, and as I fumbled with my camera, the sun glinted off its silvery case. Though the light did not hit his eyes, Manuel winced.

"In Santiago Atitlán, there are laws protecting Father Sun," he explained. "It is illegal, for example, to shine a mirror back at the Sun, because it might blind our Father's eyes. That would be an insult. There are also laws against loud sounds, screams, even loud knocking on the door at night, again out of respect for the night sky, and so as not to awaken Father Sun, who is sleeping after a hard day at work."

Sophisticates might smirk, but embedded in these laws, which are common in indigenous towns and villages throughout Central and South America, is a healthy appreciation for the power of the sky. These people sense a personal connection with the cosmos. They feel that being on good terms with the heavens enhances their life and that failure to do so would be harmful to themselves and therefore to their community as a whole.

Just off the square, we ran into Camilo, the archetypal good teacher, estimable, proud of his work and his good students, especially Manuel, whom he taught in fourth grade. I asked Camilo about 2012.

"Twenty twelve is very important. It's only four years before 2016! That's the year when the Central American Free Trade Agreement takes full effect. We must be prepared," the teacher declared.

Free trade agreement? Oh yeah, the real (irrelevant) world. Time for another ceremony, this time at Manuel's place, the Cofradia de Santiago Apostol, or St. James the Apostle. The dank room was made festive by colorful altars chockablock with statues of saints that are part animal and animals that are part divine. This place was full-blown pagan, and this former Episcopalian altar boy had a little trouble bending his knee. Manuel assisted his father in the service, a transporting litany of chants, prayers, and incense propitiations. All of a sudden Manuel instructed me to close my eyes, then spat the mouthful of rum in my face, and in Lord Byron's as well. It was a transcendent moment: part communion, part baptism, part mud-in-your-eye. After wiping off, we obeyed his command to toast the holy statues by chugging a glass of the sacred rum.

Once more I asked Manuel about 2012. He explained that it was difficult for him to work up a fear of that year because Santiago Atitlán felt so safe, "like a bird's nest, like the navel of the world." That was early August 2005.

The following month two category 4 hurricanes, Katrina and her twin Rita, pummeled the Gulf Coast of the United States. Then almost unnoticed, a category 1 pipsqueak named Stan wiggled its way across the Atlantic and hit the Yucatán peninsula, crossed over to the Gulf, and turned large portions of Central America into mud. Hurricane Stan barely made the news here because it missed the United States. It ended up being the deadliest storm of the season, killing more than 1,500 people, worse even than Katrina.

Mexico and El Salvador had many hundreds of casualties, but Guatemala was hardest hit, especially in the highlands. Virtually every river overflowed, washing out bridges, drowning livestock, fouling drinking water with sewage. Lake Atitlán was fuller than ever. The center of destruction, from mudslides, panic, and disease, was Manuel's town, Santiago Atitlán, with 650 dead, 330 missing, 4,000 homeless, and most everyone sickened, traumatized, and distraught. Panabaj, a community on Santiago's outskirts, was inundated by a wall of mud half a mile wide and up to twenty feet thick that slid down off the side of a volcano, burying all 208 residents.

Mayan funeral protocols are quite strict: Each person must be covered and buried exactly twenty-four hours after his or her death. But mud is quite heavy and gets heavier as it compresses itself with each passing hour. Pulling the bodies out of the mud proved nightmarishly hard with the bare hands and hand tools available. Helicopters ferried medicine and supplies in and out of the town square, but when President Oscar Berger's troops arrived to help with the rescue, the townspeople of Santiago refused to let them in. Their memory of the December 1990 government massacre of their loved ones was still fresh. So the bodies were left in the mud, and the village of Panabaj declared a mass grave.

Shamans like Manuel worked frantically to appease the spirits of the dead with herb and incense rituals, but a few days later their fears that the ceremonies wouldn't be sufficient were confirmed when a sharp earthquake struck. It wrecked highways and bridges and tumbled hundreds of buildings already destabilized by the rain. The region, in its death throes, was now completely cut off. Nothing and no one got in or out.

LET'S TAKE A CLOSER LOOK at the sequence of volcanic, seismic, and meteorological events that befell Central America in October 2005: On Saturday, October 1, the Llamatepec volcano in El Salvador erupted for the first time in a century, killing two people and causing thousands to flee. On Wednesday, October 5, 2005, killer Hurricane Stan made landfall on Central America and for the next four days dumped unprecedented amounts of rainfall on El Salvador, Guatemala, and southern Mexico, causing terrible flooding and landslides. On Saturday, October 8, 2005, an earthquake, 5.8 on the Richter scale, hit Guatemala and El Salvador, causing further landslides and destroying roads and bridges. All this occurred just after Hurricane Rita and just before Hurricane Wilma, the biggest storm of the year.

Are the volcano, the hurricane, and the earthquake that hit Central America unrelated events? Or are they manifestations of a larger catastrophe? Few scientists would venture an opinion one way or the other, opting to wait for all the data to be analyzed. But what hit Central America, in fact, what hit the whole Gulf of Mexico region in the autumn of 2005, was not a series of isolated, unrelated events. It was a megacatastrophe of a scale and duration rarely if ever beheld.

On what basis do I state that? On a grasp of the obvious. On the same gut feeling that inclines you to agree. Just as Elijah heard the Lord in the still, small voice that came after the earthquakes, the rending mountains, the shattering rocks, the fires, the great winds, we, in the calm that has followed the great storms of September and October 2005, have heard the Truth: There is something greater and deadlier going on here.

We have reached the point of deadly synergy, at which climatic processes communicate with and amplify each other in severe and catastrophic ways, declares Alexey Dmitriev, a renowned Russian geophysicist specializing in extreme climatology.

"As natural compensatory processes are developing (to compensate technogenic pressure on the planet), they will trigger controlling mechanisms of seismic reactions and volcanic activity, i.e., natural calamities will become more severe, up to global transformation of the climate machine and the state of biosphere," writes Dmitriev. He adds with grim amusement that one of the "advantages" of our extreme volatile position is that the "underlying mechanisms connecting seemingly diverse meteorological, seismic and volcanic phenomena" are in the process of being revealed.

The only current climatic development that could cause a megacatastrophe of volcanoes, hurricanes, and earthquakes, such as was experienced in Central America, is global warming. Warming in the Gulf of Mexico undoubtedly energized those hurricanes, in fact acted as a "veritable hurricane refueling station," according to "Are We Making Hurricanes Worse?," a special *Time* magazine cover story that explored the many ways in which human activity is fueling storm activity.

The El Salvador volcano eruption at Llamatepec was clearly part of a larger process, cooling the local climate and also contributing cloud density to the oncoming storm. The subsequent earthquake may have been triggered by the region's massive landslides, which shifted pressures on underlying faults. One tends to think of earthquakes as movements propagating upward from somewhere deep in the center of the Earth (or at least deep in the Earth's crustal layer), but sometimes, it turns out, the impetus for those movements comes from above. Faults that have been frozen in place by great weights of rock and earth bearing down upon them may suddenly become freed up when that rock and earth moves. This has led to the startling though commonsensical conclusion that global warming may lead to earthquakes, particularly in the northern latitudes. As glaciers melt, the weight they bear down on tectonic plates decreases, allowing the plates to slip around more freely. Alaska is particularly susceptible to this effect, and so therefore is the Pacific Rim.

The mechanics of megacatastrophe such as occurred throughout the 2005 hurricane season, and particularly in Central America during Hurricane Stan, are of course far from understood. And they never will be understood, given the structure of the contemporary science industry. To bring the experts on hurricanes, volcanoes, and earthquakes together to assess this situation would be a logistical nightmare, requiring radical violations of the interdisciplinary norm. Worse, it would never even occur to the jealously specialized scientific powers-that-be. Barriers of nomenclature would have to be broken. Bureaucratically unrelated professional institutions would be asked to forge links. Tenure-seekers' aspirations might even be delayed, if they were required to prepare presentations for peer groups not in their field.

The fact is that there is no scientific mechanism for examining the megacatastrophe of volcano, earthquake, and hurricane that struck Central America, or that might strike anywhere else. Yet clearly we face an emerging synergy of climatic, seismic, and volcanic threats.

Sad to say, but a wartime mentality often leads to the best science. For example, during World War II, solutions to problems were needed posthaste to fend off the enemy: curing respiratory infections plaguing bomber pilots, taking blood pressure under water, measuring infrared radiation from flash and flame. We now need to understand that we are at war again, and our new enemy is megacatastrophe. It is time for the scientific community to come together in our defense. We need their best guesses as soon as possible before the 2012 deadline expires.

At this writing, six months since the tragedies associated with Hurricane Stan befell Central America, I have been unable to contact Manuel or learn of his fate or that of his family. I keep thinking of what the young shaman said when asked what he planned to be doing in 2012: "If I am alive, I'll be doing rituals. If I am dead, there will be someone else to take my place."

THE SUN

Blaming Euclid for one's emotional problems might ordinarily seem a stretch, but not to the late shift at the old Sheridan Square bookstore in Greenwich Village. Late one Saturday night I was feeling imprisoned, I decided, by the fourth-century BCE Greek geometer's high-handed idealization of space into unrealistic three-dimensional planes. My condition had nothing to do with the fact that I had no relationship, was living alone in an overpriced, 300-square-foot roach trap which, if things didn't pick up, I might have to pay for by writing, My Peculiar Lovers, an "adult" novel that Typographical Services Inc. publishing company had invited me and my prestigious Ivy League literature degree to author. The assignment came complete with editorial guidance in the form of a mandatory dirty word list—"quim"?—and paid a total of $150, upon acceptance of the 160-page manuscript.

Darting into the bookstore just before the 2 AM closing time, I breathlessly explained my math problem to Marie, the night manager, who of course had heard it all before and who gently directed me to several dense books about a non-Euclidean geometry that embraced the reality of a spherical Earth. Curvy triangles that had more than 180 degrees. Arcs between two points that were shorter than straight lines making the same connection. What a relief!

Straightening my crumpled bills, I flashed back a decade to when I was ten, standing anxiously at the counter of Benny's Luncheonette, a Brooklyn establishment formally known but never referred to as the Park Town Café. I really, really needed a map of the Moon, like, immediately. I was desperate to escape the planet. Benny was a bit taken aback. Ice cream, cheeseburgers, newspapers, cigarettes, notebooks, tape and sundries, plenty of sundries, but no Moon maps.

"How about the Sun?" I asked.

Now that was silly. Even Benny knew that. Although the Sun was bigger than the Moon, the Earth, and all the planets put together, there were no maps of it. No Sea of Tranquility to draw, no giant red spot like Jupiter's to color in. And no one would ever go there. The Sun was just an immense ball of fire so hot that it made normal fire seem like ice and that would keep on burning forever. It would never change, and there would never be anything new about it that we would ever need to know.

6
SEE SUN. SEE SUN SPOT.

Half a mile or so past Aztec's shot-up old welcome sign, I pulled in at the Chubby Chicken. Taped right above the cash register were a bunch of sick/funny cartoons about animals getting cooked. One, captioned "Horror Movie," showed three bug-eyed chickens watching a comrade roast in the microwave. The clock radio on the order counter was playing my new country favorite, "Refried Dreams."

I was headed to Durango, Colorado, where eighty or so solar physicists from a dozen countries were gathering to explore the Sun's relationship to climate and culture. The conference was sponsored by the University of Colorado's Laboratory for Atmospheric and Space Physics, which designed and coconstructed Solar Radiation and Climate Experiment (SORCE), a research satellite that for the past several years has monitored the Sun's interaction with the Earth's atmosphere. Of the dozen or so solar physics conferences being held around the world in 2005, SORCE had the most fascinating program, with presentations on everything from the latest satellite technology to how solar fluctuations caused cannibalism in seventeenth-century China.

As my three-piece dark meat sizzled in the grease, I wondered if any more killer protons were headed our way. It was September 13, 2005, and by that point in the year I had become used to the fact that whenever I turned my attention to 2012, 2005 would jump up and, like a younger sibling desperate to be noticed, do something naughty or dangerous. On January 1, 2005, forty minutes into the New Year, Greenwich mean time, sunspot 715 released a major, X2-class solar flare. (Solar flares are rated C for light, M for middle, and X for most powerful. The numbers following the letters indicate the severity within a given class.) In and of itself, the New Year's Day eruption was certainly no cause for alarm. After all, 2005 was widely expected to be a very quiet year in terms of solar activity. But in retrospect, the New Year's Day flare set the tone for the year of Hurricane Katrina and all the other record-setting hurricanes, the year that will go down as one of the stormiest, most troubling years in the history of both the Sun and the Earth.

Perhaps there was a connection.

By every scientific measure, 2005 was supposed to have seen very few sunspots. Sunspots are larger-than-Earth magnetic storms that blemish the solar surface. They are about 1,500 degrees C cooler, and are therefore darker, than their 5,800-degree immediate surroundings. Sunspots occur in cycles of nine to thirteen years, most often eleven years, which is the usual amount of time from one solar maximum (the greatest number of sunspots) to the next. There's also an eleven-year cycle from one solar minimum to the next. It follows that the time period from solar maximum to solar minimum, from peak to trough, is usually in the five-to-six-year range. The current cycle, 23, will bottom out late in 2006. The next cycle, 24, will peak in 2012.

Sunspots have been monitored by eye for thousands of years, by telescope since shortly after Galileo invented it in 1610, and by satellite since the mid-1970s. Astronomers still have no idea why they occur in roughly regular eleven-year cycles. "It is their nature to do so," Aristotle's pat explanation for otherwise inexplicable phenomena, is about as far as we have gotten. There is however a broad scientific consensus that total solar activity, essentially meaning various forms of explosions and outbursts from the

Sun, rises as the number of sunspots rise, and falls as the number of sunspots falls.

"There is a 96 percent correspondence between sunspots and other solar activities," says Harry van Loon, a distinguished physicist now a affiliated with Colorado Research Associates and the National Center for Atmospheric Research (NCAR).

Or at least that's the way it's supposed to work. The year 2005 was the latest, and most spectacular, in a series of recent, troubling exceptions to the sunspot cycle rule. That year did see roughly the expected number of sunspots, but overall solar activity was the highest for any minimum year on record, and by some measures it was much more active than a typical maximum year.

On January 17, 2005, sunspot 720, cycle 23, a giant storm the size of Jupiter, spat out an X3-class flare. This was roughly as surprising as a spring snow flurry would be in New York City, noteworthy but hardly alarming. Sunspot 720 proceeded to erupt three more times. On January 20—a date I will always remember because forty years earlier that was the snowy day my father died in a car crash—sunspot 720 unleashed an impressive X7-class flare, more or less the equivalent of a May snowfall, say, three inches of accumulation—in half an hour.

The freakish, baffling storm shot several billion tons of protons that traveled from the Sun to the Earth in about half an hour, rather than in the usual day or two. Scientists are baffled as to how this happened. Most sunspot explosions, including the four earlier ones from sunspot 720, are of a common variety known as coronal mass ejections (CMEs). CMEs are superheated gas clouds that billow out from the Sun and plow through interplanetary space, creating shockwaves that accelerate assorted particles, mostly protons, in front of them, resulting in what's known as a proton storm. CMEs usually travel at a rate of 1,000 to 2,000 kilometers per second, pretty slow by solar system standards, and if they happen to be headed toward Earth, we feel the effects a day or two later. Satellites get zapped, certain radio communications are disrupted, and magnificent auroras fill the nighttime skies. It could be that CMEs actually play some helpful, energy-infusing role by providing beneficial stimulation to the Earth's outer atmosphere or magnetic shield. No one can say.

Sunspot 720's fifth explosion was altogether different, reaching the Earth fifty times faster than normal. If a rifle's muzzle velocity were suddenly to increase fifty-fold, the bullets coming out would be that much more powerful. Same thing with protons. The killer protons of January 20 bombarded the Earth in a freak storm that left the experts blinking.

"CMEs can account for most proton storms, but not the proton storm of January 20th," declares Robert Lin, a solar physicist at UC Berkeley.

There's just no way CME shockwaves can propel protons or other fundamental particles to such great speeds. Imagine you're in a rowboat in the middle of a pond and you throw a good-sized rock into the water, and the little bits of whatever's floating on the water are nudged forward by the ripples—that's analogous to the normal CME shockwave. To understand the January 20 event, imagine throwing a similar-sized rock really hard, so hard that it creates ripples so swift and powerful that the little floating bits now streak across the surface of the pond and smash right into the shore, shattering a few pebbles on the other side. Hard to imagine anyone throwing a rock with that much force?

This is not merely of academic interest. Light from the Sun, traveling at around 300,000 kilometers per second, takes about eight minutes to reach the Earth, meaning that, in order for the protons pushed from sunspot 720 to have reached the Earth in thirty minutes, said particles must have been traveling at, say, a quarter the speed of light, or about 75,000 kilometers per second. When anything travels at significant fractions of the speed of light, it is termed relativistic, referring to Einstein's fundamental relativity rule that matter cannot travel faster than the speed of light. Any particle, or cow or toaster, traveling at the speed of light would achieve infinite mass. At even a fraction of the speed of light, the mass gets much heavier. So those protons, instead of being nearly weightless, would have impacted the Earth with the force of tiny pebbles, quintillions of them, like a shotgun blast from the Sun. There are all sorts of caveats and counterspeculations, but unless Einstein is seriously wrong, we all will be obliterated if a future batch of protons manages to shave another twenty-two minutes off their Sun–Earth travel time, down to eight minutes, which is about how long it took for my Chubby Chicken three-piece to get nice and crispy.

An exceptional level of paranoia would be required to argue that the Sun shot the January 20 proton storm at the Earth on purpose. Old Sol has no mind or intention. If He did mean to harm us, He could no doubt find nastier ways. Nevertheless "the most intense proton storm in decades," as one NASA dispatch described it, was magnetically shepherded from sunspot 720 directly to our planet. The sunspot was said to be located at 60 degrees west solar longitude. As the Sun rotates, magnetic fields from that spot bend around to create a kind of magnetic corridor to Earth for any CMEs that erupt there.

NASA did not disclose data on the January 20 storm until mid-June; perhaps its findings were so startling that they had to recheck them. This unusual delay, as well as the puzzling lack of follow-up commentary, has made it impossible to assess the freak storm's impact, which could have been severe, since it hit Earth directly. Satellites may have been fried, skin cancers triggered. We just don't know.

The January 20 storm that opened 2005, a solar minimum year, turns out to have been the largest radiation storm since October 1989, a solar maximum year. It may well have set back plans for manned space exploration for the foreseeable future. Normally if a dangerous solar storm is headed for Earth orbit, the Moon, or wherever else astronauts might be, they have at least a day to batten down the hatches. But this hit so quickly, in less than half an hour, that the astronauts would probably not have had time to defend themselves.

Marrow-containing bones, such as those found in the skull, shoulders, spine, sternum, and thighs, are the portions of the body most vulnerable to radiation. Solar protons would obliterate the blood-producing cells living in the marrow, depleting the body's fresh blood supply in about a week.

"A bone marrow transplant would be required—stat!—but they don't do those on the Moon," writes Tony Phillips, editor of the Science@NASA dispatches.

Nor, one might add, are bone marrow transplants done on Earth, at least not in numbers anywhere near sufficient if killer protons start penetrating our planet's dwindling magnetic shield and become a health problem down here.

The year 2005 continued to be stormy, climaxing in September with one of the most turbulent weeks in recorded solar history. On September 7, sunspot 798, returning from the far side of the sun, unleashed a monster solar flare, ranked X17, the second largest ever recorded. The blast caused a blackout of many shortwave, CB, and ham radio transmissions on the daylit side of the Earth, which at the time, 1:40 p.m. EDT, included most of the Western Hemisphere. Nine more X-class flares exploded out of the Sun over the next seven days; several spurred radiation storms that pelted the Earth. The Earth's magnetic field normally protects most of us from this type of radiation. That magnetic field, however, has dwindled inexplicably in recent years.

The last solar outburst occurred on September 13, the day that the SORCE conference began. No doubt the solar physicists would be buzzing. Overall the week of September 7–13, tumultuous by the standards of any solar maximum year, was all the more astonishing because it came during a solar minimum. As meteorologist and astronomer Joe Rao of the Hayden Planetarium at the American Museum of Natural History in New York City put it, "This storm was the proverbial blizzard in July."

NOT SINCE THE ICE AGE

Sami Solanki, of the famed Max Planck Institute for Solar System Research in Katlenburg-Lindau, Germany, is a leading scientific exponent of the belief that the Sun's current behavior is exceptionally, perhaps problematically, energetic. An urbane European of Indian origin, Solanki jolted the SORCE conference: "Except possibly for a few brief peaks, the Sun is more active currently than at any time in the past 11,000 years." The physicist informed his colleagues that since 1940 the Sun has produced more sunspots, and also more flares and eruptions, which eject huge gas clouds into space, than in the past. Solanki published an earlier version of these findings in *Nature*.

If Solanki had made his announcement to a roomful of Earth scientists, a shiver of panic would have swept through the room. Eleven thousand years ago is the end of the last Ice Age, a truly iconic period. The Ice Age of 11,000 years ago, actually the latest of many ice ages that have cooled the Earth over the eons, is the greatest example of climate change we know of, in both the freezing over of much of the temperate latitudes and, as the glaciers receded and temperatures soared back to normal, the subsequent global warming. For Earth scientists, the Ice Age is pretty much the dividing point between history and prehistory.

Solanki's declaration that the Sun's peculiar behavior today is essentially unlike anything we have seen since the end of the last Ice Age is therefore no less jaw-dropping to Earth scientists than if he were to have announced to an assemblage of Bible scholars that things hadn't been like this since the time of Noah and the great Flood, which may in fact have resulted from the melting of the last Ice Age. Earth scientists had long assumed that this warming transition was, as geological processes are routinely assumed to be, slow and incremental, taking hundreds or even thousands of years. But according to a *Time* magazine special report, the latest evidence suggests otherwise: A growing number of paleoclimatologists, who study the ancient history of the Earth's climate, are reaching the conclusion that complex systems such as the atmosphere jump from one steady state to the next with only brief periods of transition, much the way that water heated to boiling suddenly turns to steam.

Richard Alley, of Penn State, specializes in the study of abrupt climate change. Alley maintains that ice cores taken from Greenland show that the last Ice Age came to an end not in "the slow creep of geological time but in the quick pop of real time, with the entire planet abruptly warming up in just three years. Most of the time, climate responds as if it's being controlled by a dial, but occasionally it acts as if it's controlled by a switch."

Is the Sun about to flip the Earth's switch? For the record, Solanki made no pronouncements about any effects the Sun's current behavior might have on the Earth's climate. He simply observed that the Sun appears to be more active today than at any time since the end of the last Ice Age. If such pervasive and dramatic climate change were to occur again today, with 6 billion more people living on the planet, bound together in an interdependent global economy, the results would be catastrophic far beyond anything in human history or imagination. Especially if such a change were to happen, as Richard Alley contends, in a "quick pop of real time," say, between now and 2012.

THE HOTTEST IT HAS BEEN IN 50,000+ YEARS

Virtually all the data regarding climate history back to the Ice Age is from ice core samples taken from the Arctic or Antarctic. But extrapolating the climate history of equatorial regions from polar ice samples is a pretty chancy affair. Imagine having weather data from only the northernmost and southernmost thirds of the Earth and trying to figure out what went on in the middle— 11,000 years ago. This is particularly problematic since approximately two-thirds of the world's population, including the Mayan descendants, live closer to the equator than to the poles and therefore in zones relatively untouched by Ice Age glaciation. But where are you going to get ice at the equator?

The answer, according to Ohio State University glaciologist Lonnie

Thompson, one of the most celebrated scientists of our era, is from way above sea level, 3 miles high at least.

Though he's always invited, Thompson doesn't make it to many conferences, including the 2005 SORCE. Instead he's usually climbing mountains, which in point of fact he doesn't much like to do; he has asthma and would prefer staying at home in Columbus, Ohio, with his wife and coresearcher, Ellen Mosley-Thompson. But that hasn't kept him from spending more time than any other human being at altitudes of 18,000 feet and higher (Sherpa guides included, commercial pilots not). For the past thirty years, Thompson has simply followed his eminently commonsensical observation, that one cannot deduce tropical climate from polar data, to its ultimate logical conclusion—scaling the mountains closest to the equator for clues to that region's climate history.

Thompson and his team have assembled a library of ice samples some 4 miles long, stored at the Ohio State campus in a 2,000-square-foot refrigerator facility that maintains the cores at temperatures of -30 to -35 degrees C (-22 to -31 degrees F). Ice cores are literally frozen pieces of time. The deeper the ice core, the more ancient the history. By analyzing the chemical content of each layer of ice, researchers determine a timeline of the climate in the locality from which the ice core was drawn. Thousands of such ice cores have been analyzed, creating a database that has allowed researchers to gradually piece together the planet's climate story back to the Ice Age, and in some cases much further.

Like the H-bomb-proof vault in Reykjavik, Iceland, which holds the scrolls bearing the great Icelandic sagas, the Ohio State ice core refrigerator preserves history that can never be replaced. Thompson's facility should be declared a world heritage site. Indeed global warming has been melting the world's glaciers at an accelerating rate, giving Thompson's team even more impetus to sample the world before our heritage is lost.

"Global warming is not as controversial as some people would like you to think. The evidence is clear that a major climate change is under way," says Thompson. Hailed with a raft of awards, consulted by Al Gore, *National Geographic,* and the *New York Times,* Thompson is the basis for the character played by Dennis Quaid in *The Day After Tomorrow,* a global-warming thriller. Thompson, who was climbing a peak in China at the time of the SORCE conference, is best known for his conclusion that Kilimanjaro, the African peak that Hemingway made famous for its snows, is in fact losing its ice cap and will do so entirely by 2015.

When asked of the consequences, Thompson's response is touchingly human: "Tourism is the biggest single industry in Kenya, and it will probably drop off if Kilimanjaro's famous snow-capped peak isn't there any more."

For years Thompson has been assembling a mass of evidence to the effect that 5,200 years ago the Earth experienced a climate catastrophe. Citing studies on everything from tree rings to human corpses, from plant pollen to oxygen isotopes, he concludes that 5,200 years ago a sharp drop, then a surge, in solar activity transformed the Sahara from a greenbelt to a desert, shrinking ice caps at the poles and otherwise disrupting and distressing global ecology.

It is interesting to note that this 5,200-year period coincides with the Mayan definition of an Age, or Sun. Recall the Barrios brothers' explanation that we are now in the Fourth Age, which began in 3114 BCE and will end in 2012. Indeed, 3100 BCE seems to have been a pivotal time in many regions. That is when ancient Egyptian civilization first arose and also when, in Hindu mythology, Krishna died and the current age, the Kali Yuga, or Degenerate Age, began. It could well be that the end of the global ecological crisis that occurred 5,200 years ago marked the rise of new civilizations and the beginning of a new era.

Thompson believes that the conditions that led to the disaster of 5,200 years ago are very similar to what we are experiencing today. "Something happened back at this time and it was monumental. But it didn't seem monumental to humans then because there were only approximately 250 million people occupying the planet, compared to the 6.4 billion we now have. The evidence clearly points back to this point in history and to some event that occurred. It also points to similar changes occurring in today's climate as well," he warns.

Thompson regards mountain glaciers such as Kilimanjaro as the Earth's "crown jewels." Their loss, and the loss of water they provide to the land below, will inevitably lead to drought, famine, and a shortage of hydroelectric power—in short, to catastrophe for societies that depend on that water and eventually for the regional and global communities of which they are a part.

In retrospect (if there is one), today's climate may make the situation 5,200 years ago seem like a walk in the park. On a stroll through a meadow on one of his favorite glaciers, Peru's Quelccaya ice cap (shrinking forty times faster than when first studied in 1963), Thompson happened upon some unusual plant fossils, so he packed up some specimens and shipped them off to two independent laboratories. The tests came back showing that these specimens were between 48,000 and 55,000 years old. For the plants to have been in the near-perfect shape they were found in, they had to have been covered and protected by ice for most of that time, "which means that the ice cap most likely has not deteriorated to its current size for any length of time in more than 50,000 years," according to Thompson.

So it's hotter now than it has been in 50,000 years or more. Perhaps the figure is more like 74,000 years, when the Lake Toba supervolcano spewed ash into the atmosphere, which made the air unbreathable, blocked the Sun's light and heat, led to an ice age, and decimated humanity.

DATING RITUALS

Like the kid in the back of the class who's had his hand up so long that he's using the other hand to hold it up, the reader might be wondering: how can they be so sure of what happened thousands and thousands of years ago, before any sort of records were kept?

The answer, in a nutshell, is carbon 14. Carbon 14 is a radioactive isotope that has six protons and eight neutrons, two more neutrons than the regular element carbon, which, with six protons and six neutrons, adds up to an atomic weight of 12. There's so much carbon from plant and animal organic matter in the world that carbon 14 can be found almost anywhere, in strictly predictable proportions to a sample's overall carbon content. This isotope begins to decay the moment a plant or animal dies, and its half-life, the time it takes for half of a given amount of carbon 14 to become nonradioactive, is 5,730 years. Mass spectrometers can now literally count the number of carbon 14 atoms, enabling precise datings to be made from very small samples.

Carbon 14 is made radioactive by cosmic rays from outer space impaling its nucleus. It turns out that there is an inverse relationship between sunspots and the number of cosmic rays that make it to Earth—the more sunspots there are, the denser the interplanetary magnetic field emanating from the Sun, and therefore the fewer cosmic rays able to make it to Earth and bombard stable carbon to render it radioactive. The same holds for the element beryllium, another radioactive isotope used to establish dates. The more sunspots there are, the less radioactive beryllium 10 is created.

Tracing the Sun's behavior back to pretelescope days requires investigating residual evidence of sunspots, thereby deducing their number and intensity. Sunspot activity can thus be inferred by assessing carbon 14 and beryllium 10 levels at different points in history. Back to the ice core samples. The general rule is that the deeper down in the ice, or the ancient tree trunk, the earlier the isotope was deposited.

Without carbon 14 and other radioactive isotope dating techniques, we would have no knowledge about solar activity prior to the invention of the telescope in 1610. That would be a grievous intellectual loss, since four hundred years is woefully insufficient for understanding long-term climate trends on our 5-billion-year-old planet. Without historical context, it is impossible to assess significance—it is impossible to tell whether the Sun is truly misbehaving or just going through the same sort of phase it has gone through many times before.

For example, the century and a half from 1100 to 1250 was unusually warm here on Earth. During that time, the Vikings were able to establish flourishing colonies in Greenland and even in northeastern Canada, which they dubbed Vinland, for the wine grapes that apparently grew there. Carbon 14 records clearly indicate that the Viking heyday was also an era of unusually high solar activity. There is general agreement, however, that the solar activity during the Viking times was far less than today's. If back then the Sun could be said to have developed a case of sunspot acne, today it's breaking out in hives.

Much of the SORCE conference was devoted to debating how accurate isotope dating techniques really are. Are levels of carbon 14, for example, affected by anything other than cosmic radiation? What's the better indicator, absolute levels of radioactive isotopes or the rate of increase/decrease? How accurate are radioactive isotope measurement techniques anyway? Do fluctuations in the Earth's magnetic field influence isotope readings? (That was a hot topic.)

On balance, the use of carbon 14 to infer historical sunspot activity withstood the onslaught of the SORCE conference skeptics, though with certain provisos, such as distortions caused by nuclear weapons testing, which creates carbon 14, and fluctuations in the Earth's magnetic field, which can independently affect the number of cosmic rays penetrating the atmosphere. Beryllium 10 was deemed somewhat less reliable because of its tendency to attach itself to aerosols, which float around in the atmosphere for a year or two and then deposit themselves haphazardly. So one tree ring containing very little beryllium 10, which would indicate a high level of sunspot activity, might be misleading, because another tree ring from the same historical era might contain a great deal of beryllium 10, simply because the second tree was more efficient at absorbing aerosols.

Prudence, all agreed, dictates that work relying on measurements of either isotope should be checked more carefully in the future.

"If there is a future!" I wanted to shout, but that would have been way uncool.

THE PHYSICISTS' NEW CLOTHES

I felt like the little boy pointing to the naked emperor, except that this time the emperor was on fire. As noted, the seven-day period from September 7 to 13, 2005, was one of the most tumultuous weeks in the known history of the Sun, really scorched the record books, but at the SORCE solar physics conference, which began on September 13, there was barely a mention of the storm.

For the record, this information was fully available during the SORCE conference. I know, because I checked NASA dispatches every day by e-mail and found headlines such as "Intense Solar Activity," "Ruby Red Auroras in Arizona," and then "Solar Minimum Explodes!" written by Tony Phillips. Several weeks later, on September 26, the Sun protruded its largest prominence in recent memory. The limb-shaped fireball was many times the size of the Earth. Overall September 2005 turned out to be the most tumultuous month on the Sun since March 1991—which was a solar maximum year and was therefore expected to be turbulent.

In the annals of solar physics, September 2005 is destined to take its place beside the now legendary series of solar upheavals known as the Halloween storms, which took place between October 26 and November 4, 2003. For the first time in astronomers' memory, two Jupiter-sized sunspots appeared on the Sun's face at the same time. Both then proceeded to explode over and over again with X-class flares. The storm began on October 26 and climaxed on November 4 with the largest solar flare ever recorded, an X45-

class haymaker. Had the resulting coronal mass ejection headed toward Earth, it would have flattened the global satellite network. Telecommunications, banking, and even military surveillance satellites would certainly have been fried. We know this because a smaller X19 flare issued a radiation storm that hit the Earth in 1989, knocking out the Hydro-Quebec power grid for several hours, fusing some generators solid. The human health consequences from an X28 flare's storm, in the form of radiation poisoning, cancers, eye maladies, and other disorders, might well have been severe.

Countless articles, blogs, and commentaries about the Halloween 2003 storm swirl around the Internet, many of them hysterical and confusing but nonetheless usually grasping two main points: (1) this storm period was extraordinary in its ferocity, and (2) had the fallout hit Earth head-on, we really might have suffered. Halloween 2003 was so powerful that some solar physicists now refer to it as a second solar maximum, since it came two to two and a half years after the solar maximum of 2001, and also because the Sun's behavior never really settled back down to normal. September 2005, though slightly less powerful than Halloween 2003, was even more significant, because it came at the trough of the cycle.

So why, at a SORCE conference, organized by those who operate a solar research satellite, was there barely a mention of what will go down as one of the most remarkable weeks in recorded solar history? True, the SORCE newsletter did several months later feature the September 2005 storm, but why no brainstorming while all the solar physicists were gathered together in one place to share ideas?

Solanki gently explained that most scientists only get excited when all the data are in. The September 2005 events would likely be the hubbub of next year's SORCE conference, or the one after that. Intellectually I could accept the scientists' painstaking, wait-and-see methodology, but not emotionally. Their workaday nonchalance was an eerie denial of real-time emotional response to the spectacular, unprecedented doings on the Sun, the central object of every solar physicist's professional life, happening right then and there in the heavens above them.

Hell, I couldn't even accept it intellectually.

September 2005 was turning out to be one of the stormiest, craziest months in the history of the Sun and the Earth. The overheated waters of the Atlantic and the Gulf of Mexico just could not let off enough steam. Katrina, the immortal one, had already destroyed New Orleans (Sodom and/or Gomorrah, as per some of those biblically inclined). Rita scared Houston and Bush and dumped a lot of rain. At the beginning of October came Hurricane Stan, that little-known Central American hurricane that devastated Atitlán and turned out to be the deadliest killer of the year. Then Wilma, the most powerful of the bunch, shorted out Florida. At least eight other tropical storms, some of hurricane strength, followed, for a total that far outstripped any other season.

The year 2005 was well on its way to becoming the hottest, stormiest, and yet also the driest year ever recorded. It may well have been one of the most seismically and volcanically active as well. It even ended with a highly unseasonable spate of grass fires and tornadoes, hardly a holiday tradition.

Even the SORCE handbook specified a connection between solar activity and the climate on Earth: "Energy balance equations predict that if the Sun varies by a modest amount, say 1%, the global average surface temperature will change by about 0.7 degrees C. Some empirical models estimate that the Sun has varied by nearly 0.5% since pre-industrial times. Climate models indicate such a change may account for over 30% of the warming that has occurred since 1850," according to the SORCE handbook they passed out at the door.

Over 30 percent? That would make the strengthening Sun more important to global warming than any factor except the fabled increase in CO_2. So it seemed eminently reasonable to investigate the connections between the solar and terrestrial tumult that was unfolding by the minute. Yet here were eighty accomplished solar scientists, together for three full days, and not so much as a coffee break devoted to exploring this frightening coincidence.

NEVER PREDICT!

If storm periods on the order of September 2005 are happening right near the solar minimum, what, pray tell, does the next solar maximum, in 2012, have in store for the Sun and Earth?

"Never predict!" intoned veteran researcher Harry van Loon, after his masterful presentation correlating sunspot variability with precipitation patterns around North America. But as Richard Feynman, the legendary physicist, long maintained, the most significant ability of science is its ability to foresee. We need our solar physicists to make predictions, to best-case and worst-case some 2012 scenarios for us.

If the solar maximum period starting in 2011 and peaking in 2012 turns out to be as far above the average solar maximum as the period from Halloween 2003 to September 2005 was above the average solar minimum, then we may indeed be in for the catastrophe Mayan astronomers have been warning us about for the past 1,500 years.

Several months after the SORCE conference wrapped, a team of solar scientists from the National Center for Atmospheric Research (NCAR) in Boulder, Colorado, confirmed what so many had come to suspect: "We predict the next solar cycle will be 30 to 50 percent stronger than the last," said Mausumi Dikpati. Along with Peter Gilman and Giuliana de Toma, also of NCAR's High Altitude Observatory, Dikpati has developed the predictive flux-transport dynamo model, which generates solar activity forecasts by tracking the subsurface movements of sunspot remnants from the two previous cycles. Based on new helioseismology techniques, in which sound waves inside the Sun are tracked much as a physician might use ultrasound to see inside a human patient, the NCAR team believes that sunspots help beget more sunspots in a convoluted, conveyor-belt-like process. "When these sunspots decay, they imprint the moving plasma with a type of magnetic signature," Dikpati observes.

Sunspots start out as magnetic knots in the solar convection zone, which is the outermost layer of the Sun's body and also the region likeliest to be disturbed by external electromagnetic or gravitational influences. Currents of plasma, or highly electrified gas, act as conveyor belts and ferry these knots from the poles to the equator, where they rise up to the surface and explode as magnetic storms, what we call sunspots.

"Predicting the Sun's cycles accurately, years in advance, will help societies plan for active bouts of solar storms, which can slow satellite orbits, disrupt communications, and bring down power systems," declares the NCAR communiqué.

The NCAR team's findings, published in the prestigious *Geophysical Review Letters,* indicate that the next solar cycle, cycle 24, will start in 2007, six months to a year later than expected. It will be 30 to 50 percent stronger than this last, record-setting cycle, and will climax in 2012 . . .

On his last day on Earth, Elijah was caught up in a whirlwind and then was borne up to Heaven in a chariot of fire drawn by a horse of fire. May he choose to return, in just the same manner, at the next SORCE conference.

7

AFRICA CRACKING, EUROPE NEXT

Jah must be angry. Why else would Rasta Cabbie's Almighty put asunder the sacred, ancient homeland of His beloved prophet, the Imperial Majesty Haile Selassie, emperor of Ethiopia, Lion of Judah, also known as Ras Tafari, a diminutive man who died in 1975 but who nonetheless remains a towering figure, a living messiah in the line of Moses, Elijah, and Jesus?

On September 14, 2005, the day after the last of ten X-class flares billowed out from the Sun into the Solar System, making the seven-day span of September 7 to 13 one of the most turbulent weeks in recorded solar history, an earthquake in remote and desolate Boina, Ethiopia, about 270 miles northeast of the capital, Addis Ababa, split open a crack in the Earth 37 miles long, according to an Associated Press report. Over the next three weeks, the fissure in Boina widened to a gap of 13 feet, and it continues to spread today. Researchers from Ethiopia, Britain, France, Italy, and the United States believe that this fissure is literally the beginning of the process of the continent of Africa cracking apart into two or more pieces.

"We believe we have seen the birth of a new ocean basin," said Dereje Ayalew, from the University of Addis Ababa. Ayalew leads the eighteen-member multinational research team monitoring Boina. He presented its findings to an American Geophysical Union (AGU) meeting in San Francisco in December 2005. "This is unprecedented in scientific history because we usually see the split after it has happened. But here we are, watching the phenomenon." The research team believes that, at the present rate of spreading, it will take about 1 million years for a new ocean to form and fill in with water. (For purposes of comparison, 1 million years, in the life of the 5-billion-year-old Earth, is the proportional equivalent of about five or six days to the average person.) Of course further earthquakes could speed up the process considerably.

The cracking apart of his spiritual homeland will no doubt lead Rasta Cabbie to implore Almighty Jah, though I'm not sure if he will pray for the process to cease or accelerate. Funny that a number of survivalists are reported to have chosen Ethiopia as the place to ride out 2012. Scuttlebutt has it that that's where Robert Bast, the Australian doomsday enthusiast who runs the Dire Gnosis Web site devoted to trumpeting the upcoming 2012 calamity, is staking his claim. Keep an eye on that Boina crack, is all I have to say.

Is there any relationship between September 2005's extreme solar activity and the subsequent megacrack in the Earth's crust?

When enough electrical energy collects in the atmosphere, it is sucked down as lightning and is conducted beneath the Earth's surface. Areas rich in iron and other metallic ore deposits conduct this electricity out of the atmosphere and into the ground, thus helping to stabilize the climate. The Bermuda Triangle, for example, is believed to be densely populated by iron-rich underwater conductors. For the most part, this energy input dissipates harmlessly, but occasionally extra large bursts of energy, such as might come from extreme solar activity, may well have volcanic or seismic consequences, such as the Biona earthquake, perhaps. No one knows for sure.

Suppose someone did know. Suppose a competent team of researchers discovered that the seven days of unusual solar activity, from September 7 to 13, 2005, caused and/or contributed to the September 14 Boina earthquake and to the eventual splitting of the African continent. Would these findings reach the light of day? Should they? Is there a global censorship mechanism, a covert oligarchy that suppresses such potentially volatile news? For the record, I have no present knowledge of any such cabal, as evidenced by the fact that you are able to read this book. Although I could certainly see the rationale for suppression—

to preserve social stability. Maybe they'll call me a nut, and I'll be discredited by research organizations around the planet. That could be their tactic.

Were the public to sense a Sun-seismic connection, the next solar flare-up could cause quite a panic. "AFRICA CRACKS! EUROPE NEXT!"

HURRICANE RUNWAY

Boina, at approximately 11.25 degrees north latitude, is just off the south-eastern tip of the Sahel savannah strip that crosses north central Africa, separating the Sahara desert above it from the tropics below. This quasigreen-belt runs between 11 and 20 degrees north of the Equator (virtually the same latitudes that contain Mayan territory). From Africa's east coast, right about where the Red Sea feeds into the Indian Ocean, the Sahel crosses all the way west to Senegal's Atlantic coast, the spot, it turns out, where all Atlantic hurricanes are born.

"All Atlantic hurricanes, no matter how grand they may become, begin the same. Each starts as a disturbance in the atmosphere above equatorial Africa. These disturbances, called tropical waves, head west and, if conditions are just right, they increase in size and start spinning. Some develop into depressions, grow into tropical storms and finally evolve into full-blown hurricanes," reads a NASA dispatch.

What causes the hurricanes to form off Africa's west coast in the first place? There are two complementary theories: (1) that rainfall, particularly thunderstorms in the Sahel, creates the tropical waves, which turn into the tropical depressions, which on occasion eventually upgrade into hurricanes, and (2) that rainfall, particularly thunderstorms in the Sahel, prevents desert winds from damping down tropical depressions and keeping them from becoming hurricanes.

Both theories agree that Katrina, Rita, Andrew, Hugo, Camille, and on and on—all our hurricanes—are now believed to have started as thunderstorms in the western Sahel. The lag time between the Sahel thunderstorms and North American landfall of the hurricanes they ultimately generate runs a week to ten days.

But there are thunderstorms every day, all over the world, and the vast majority of them do not turn into hurricanes. Something extra is going on in north central Africa, some extra input, perhaps a "butterfly effect," the extra little push that occurs in the right place at the right time to start the whole thing rolling, as per chaos theory. Exactly what that first little push might be—desert whirlwinds in the atmosphere, shockwaves funneled into the eastern Sahel by Indian Ocean monsoons—scientists cannot yet say. However, they are in agreement that the impetus comes from somewhere along the Sahel ecological continuum.

In the 1970s the Sahel descended into its worst drought in modern history, from which it began emerging several years ago. The return of the Sahel rains coincided almost exactly with the spike in sunspot activity from Halloween 2003 through September 2005 and beyond. The years 2004 and particularly 2005 were the rainiest the Sahel had seen in quite some time, leading, as the theory goes, to two of the most intense Atlantic hurricane seasons in history. In the aftermath of Katrina, it is easy to forget that the 2004 hurricane season, with four whoppers hitting Florida one right after the other, was almost as bad as 2005.

So, right after a week of historic sunpot activity, the eastern part of the Sahel cracked apart, just as record-breaking hurricane activity peaked along the western part of the ecological continuum. Coincidence, or catastrophic synergy?

THE FACT THAT AFRICA began to crack apart at the height of all this may well be more than just a coincidence. If the Sahel's west coast finds itself in unprecedented tumult, the east coast of this ecological continuum might logically be affected as well.

The relationship of sunspots and other solar outbursts to thunderstorms, hurricanes, volcanoes, and seismic events here on Earth is exactly the kind of question that should be addressed during the International Heliophysical Year (IHY) 2007, a twelve-month global program of symposia and research initiatives that will promote the study of the Sun. IHY 2007 is the fourth in a series of international scientific research years, the most recent one being the International Geophysical Year (IGY) of 1957–58, which bolstered earth sciences and which spurred the Soviet Union into its October 1957 Sputnik launch to celebrate the event. Earlier such research years were the International Polar Year of 1932 (South Pole) and the International

Polar Year of 1882 (North Pole). All such international years proceeded without significant political incident.

No protests have thus far been announced for IHY 2007, but don't be surprised if this time there arises a populist demand for more complete disclosure of solar activity data—data vital to our personal and ecological health, which has been gathered almost exclusively with public funds. The NCAR model predicting a record solar maximum for 2012 will almost certainly come under attack, from solar physicists caught flatfooted upholding the status quo and also, I believe, from researchers with even more dire predictions for the coming cycle. If the NCAR team's work is not prominently featured, we will have witnessed the triumph of politics over science and over the commonweal. Fear of controversy over the dangers of the coming solar climax will have subverted the scientific community's duty to help us plan and prepare. We, the world's taxpayers, do have some leverage on the solar physics establishment, including ultimate veto power over the numerous expensive solar satellite proposals that will undoubtedly be debuted at IHY 2007. We have, after all, financed quite a fleet. Starting in the mid-1970s, when Helios I & II first went aloft, more than a score of solar research satellites have been launched, mostly by NASA and ESA (European Space Agency). In 1980 the Solar Maximum Mission was sent specifically to monitor solar activity at the apex of the sunspot cycle. In 1990 the joint NASA-ESA Ulysses focused on specific parts of the solar spectrum, such as X-rays, visible light, and ultraviolet, as did Japan's Yokoh Solar A satellite in 1991.

The current satellite generation examines solar events that particularly affect Earth. The greatest of all solar probes, SOHO (Solar and Heliospheric Observatory), launched in 1995 and still going strong, has the mission of identifying earthbound coronal mass ejections, solar flares and the like, and of warning scientists far enough in advance so that they can defend satellites, power grids, and other Sun-sensitive technologies with shielding mechanisms. The dirty little secret of the global satellite industry is that many of these satellites, particularly commercial ones, are unprotected from potential solar outbursts. Solar flare shielding is expensive and cumbersome, limiting satellite functionality. Such cost-benefit assessments would normally be the province of the companies that own the satellites, except for the fact that an increasing share of the military and intelligence traffic is handled by unprotected commercial satellites. Thus a series of massive solar storms, such as those expected for 2012, could not only knock out commercial telecommunications but could disable key military systems as well.

TRACE (Transition Region and Corona Explorer) launched in 1998 examines the magnetic structures, including sunspots, that appear on the Sun's surface. And RHESSI (Reuven-Ramaty high energy solar spectroscopic imager) has provided X-ray and gamma-ray images of solar flares since 2002. The SORCE satellite, operated since 2003 by the Laser and Spectrum Physics (LASP) Laboratory of the University of Colorado, has the mission of exploring solar effects on the Earth's atmosphere. Late in 2006 NASA will launch STEREO, a pair of satellites that will in effect act like a pair of eyes providing three-dimensional views of coronal mass ejections. Also launching in 2006 is the Yokoh Satellite B, which will provide very-high-resolution images of solar events. Beginning in 2008 NASA's Solar Dynamics Observatory will study the impact of solar events on the Earth.

All in all quite an armada to study the Sun, supposedly the very essence of stability. Would so much time, money, and talent have been invested in studying the Sun if the interest were purely academic? Perhaps it is time for the scientific and military powers-that-be to come clean about the fears and motivations behind such a massive research undertaking.

LITTLE ICE AGES

Science is not without its politics, or its embarrassments. David Hathaway, a NASA solar physicist who more than anyone else has voiced the status quo party line about there being nothing particularly unusual in recent sunspot activity, looked kind of lonely when the 2006 NCAR report on solar cycle 24 came out, predicting a massive climax in 2012. Hathaway, a respected and passionate scientist, graciously bowed to the NCAR report. He nonetheless raised some eyebrows when, several weeks later, he produced the hypothesis that the following sunspot cycle, solar cycle 25, projected to climax in 2022 or so, would fall far below average. Never say die.

The comfort I took in Hathaway's low-solar-activity prediction began to erode when I remembered what Gerardo Barrios had to say regarding the Earth-Sun relationship. Barrios observed simply that, like any other relationship, imbalance was the threat. Too many sunspots, too few sunspots—trouble can come from either extreme.

God save us from another Maunder Minimum, a seven-decade period from 1645 to 1715, when sunspots nearly flatlined, with only forty to fifty observed telescopically, during a time frame that would normally have seen hundreds if not thousands of eruptions. The Maunder Minimum is believed to have been caused by the Sun expanding in volume, and therefore decreasing in density, and also by a slowing of its rotation. The result was a less energetic, less emissive Sun, which threw off less heat.

The Maunder Minimum coincides with the heart of what on Earth has come to be known as the Little Ice Age, which appears to have begun around 1300 CE, when summers in Europe became unreliable, with too few warm, sunny days to sustain the crops. Then came the Great Famine of 1315–17, when rains drenched Europe's spring, summer, and fall, preventing grains from ripening in the fields. More than a million starved, leading, among other things, to the wholesale abandonment of children, as told in the story of Hansel and Gretel.

Winters throughout the Northern Hemisphere grew progressively colder, reaching their bitterest in the mid-seventeenth century, the time of the Maunder Minimum. In Switzerland, glaciers in the Alps advanced. In the Netherlands, canals and rivers froze over. The former Viking colony in Iceland lost half its population; the colony in Greenland died out entirely. In Africa, snow was reported in many regions where it is not seen today. Timbuktu, the ancient university city of Ethiopia, was flooded many times, though there are no records of that happening before or since.

In continental Europe, mounting political tensions resulting from the harsh climatic conditions took form as the Thirty Years War, 1618–48. In Germany, mortalities from starvation, warfare, and disease reached 15 to 20 percent of the population. England was destabilized by two civil wars, known as the Puritan Revolution, or the Great Rebellion. No wonder that this is when North America began to be colonized. Religious persecution? How about the threat of mass starvation? You have to be pretty damn desperate to jump on a rickety little wooden ship and head out on the high seas. That first Thanksgiving was gratitude for having finally found a good meal.

Moralists among us might flinch at blaming a society's wholesale descent into cannibalism on dysfunctional sunspots, but according to Sultan Hameed, a SUNY Stony Brook solar physicist who gave one of the most exciting presentations at the SORCE conference, the Maunder Minimum closely correlates with the decline and fall of the Ming dynasty in China. Drawing on meticulous records compiled by 2,000 years' worth of memos penned by Chinese civil servants, Hameed methodically demonstrated that from 1628 to 1643 China suffered fifteen years of severe drought; in the past it had taken only three years of such drought to lead to starvation. Famine, disease, outbreaks of locusts, and eventually widespread cannibalism precipitated spontaneous uprisings in different parts of the country, which led in 1645 to the overthrow of the Mings by the Manchus, who went on to form the Qing dynasty, which ruled until 1911.

Imagine if today's China, with its 1.5 billion people, quickly rising to become the world's leading economic power, were once again faced with fifteen years of drought. China would fall into chaos, with geopolitical fallout rippling throughout the rest of the world. A wounded superpower is a dangerous thing. The last insurrectionary era in China, in the mid-twentieth century, when the Communists took over under Mao Tse-tung, left at least 20 million dead. Without China's stabilizing influence, both North Korea and Iran might become bolder and more bellicose. And the global consumer marketplace would suffer mightily if the flow of inexpensive Chinese goods were disrupted; Wal-Mart, the largest company in the world, would lose its greatest single source of products.

WHEN JUPITER ALIGNS WITH MARS

It's only natural to want to discredit terrifying news. I had set out for the SORCE conference in Durango, Colorado, to find out if there is any connection between tempests on the Sun and tempests on the Earth. Clearly there is, and just as clearly we are headed for even greater tumult between now and 2012. My assumption, of course, is that what happens on the Sun causes what happens on the Earth, and not vice versa. Lingering questions as to why the great hurricanes of 2005 both preceded and followed the terrible Sun storm week of September 7–13 would, I took it on faith, be dispelled one day. I subsequently learned that there is a promising body of scientific research holding that the planets, including the Earth, help cause sunspots as well as being affected by them. It turns out that planetary configurations and alignments have a powerful influence on the Sun. This realization prompted what might be called an out-of-body experience, or at least the vivid remembrance of a very pleasant one from long ago and far away.

Nothing could have made a teenage boy more optimistic. My astrological sign is Aquarius, and for my fourteenth birthday I got to see *Hair* on Broadway. In the closing number before intermission, the entire cast stripped and sang "The Age of Aquarius." Two dozen naked people, half of them pretty girls, proclaimed that it was the dawning of my age.

At first, the science of planetary configurations and their energetic effects upon the Solar System was almost impossible to take seriously, because the music would not stop playing in my head: "When the Moon is in the seventh house, / And Jupiter aligns with Mars . . ."

Astrologers start from the assumption that planetary alignments have significance, a position I had dismissed as unscientific until researching this book. Before my investigation into 2012, I thought of astrology as sincere and entertaining but largely unworthy of serious consideration. True, one cannot help, at times, being involuntarily impressed by how certain types of personalities do seem to correspond, beyond random probabilities, with certain birth signs. And a competent reading of one's astrological chart (ever so absorbing because it's all about oneself) can reveal past and future events, as well as hidden present conditions, to a notable degree. But I had always assumed, without giving it much thought, that all that planetary stuff was just a vehicle through which certain genuinely intuitive, perceptive individuals— good astrologers—somehow channeled their perceptions.

But it turns out that there is genuine scientific value to the study of planetary configurations, maybe a lot. A devoted cadre of space scientists now believe that the planets regularly exercise significant, and heretofore largely unappreciated, electromagnetic and gravitational influence on the Sun. At first, common sense rejects this suggestion: How could such comparatively tiny, inert orbs impact the mammoth, radiant Sun they circle? But then we recall that the Sun is, unlike the planets, liquid and squishy. Like molten jello, it is far more susceptible to magnetic and gravitational pulls and tugs.

Mercury, Venus, Earth, and Mars are considered the inner planets, being on the Sun side of the great asteroid-filled gulf that separates Mars and Jupiter. Of these, Earth has the largest mass, the strongest gravitational field and also by far the strongest magnetic field. The Sun-Earth connection, therefore, is a two-way street.

That an energetic feedback system exists between Sun and Earth raises some interesting possibilities. Hurricanes, volcanoes, earthquakes, and other climatic/seismic events in which large amounts of energy are released could both cause and be caused by sunspots. More important than any of the details is the shift in perspective, from one-way transmission from Sun to Earth, to a two-way (if still lopsided) energy relationship. Indigenous ceremonies, such as the ones Manuel the Mayan shaman performs, ritually acknowledge the Earth's influence on the Sun, and they have done so for millennia.

TAKES TWO TO TANGO

It takes two, or in this case, twelve—Sun, ten planets (including the new Planet X), and the Earth's Moon, which is one of the largest moons in the solar system and therefore is a significant gravitational factor—to tango.

The Vital Vastness, a meticulously referenced, 1,000-page scholarly tome that has become something of a cult classic among geoscientists, summarizes the scholarship on how the planets, particularly the Earth, electromagnetically and gravitationally affect solar behavior. Just as astrologers calculate the angles between planets to determine their relative influence, so do author Richard Michael Pasichnyk and the other space scientists who share these beliefs. The greatest combined influence may come when planets line up (0 degrees angle between them), or when they are opposite each other (180 degrees), or even when they are square (90 degrees). Some configurations, for example, are more effective in creating rifts in the Sun's outer layer; still others apparently are best at yanking Old Sol's bowels.

"The Earth's magnetic field undergoes changes of intensity that reflect the magnitude of changes in solar activity *before* they take place on the Sun . . . magnetic data for the Earth at sunspot minimum indicates the 'depth' of the following maximum," declares Pasichnyk (italics his). In other words, developments in the Earth's magnetic field precede, and presumably help cause, developments on the Sun.

It is interesting to note that the period of great hurricanes in 2005 bracketed the record solar activity week of September 7–13. Katrina shortly preceded the flare-ups, and then Rita, Stan, and Wilma followed the sunspot outburst almost immediately.

The so-called primitive cultures that personify the Sun would not have much trouble understanding this dynamic. Their beliefs, however mystically derived, hold that the Earth and Sun are in a relationship, meaning that each influences the other, for good and for ill.

Planets and stars are giant magnets, among many other things. To understand how they interact energetically with each other, imagine taking two magnets, one in each hand. First, spread your hands apart so that the magnets are far enough away from each other that there's no pull between them—no magnetic interaction. Now slowly bring your hands together. At a certain point you will feel a force, either attraction or repulsion, depending on how the magnets' poles are oriented. Spin the magnets round and round, and electricity (minute amounts in this case) will be generated between them. Different angles and positions, in fact, create different electromagnetic fields, of varying characteristics and intensity. That's what the interaction between two planets is like. Now add a third magnet, say a million times larger and more powerful than the ones in your hands, sitting like a great white-hot blob of gelatin in the middle of the room. This immense magnet is analogous of course to the Sun, and it will have powerful electromagnetic relationships with each of the magnets in your hands.

What we tend to forget, however, is that in their way the little magnets in your hands are influencing the gelatinous mass in the middle of the room. Even though the giant blob is far more powerful energetically than those little magnets, it is gelatinous. Its surface and interior are susceptible to even the slightest disturbance.

Back to the magnets in your hands. Spread your hands apart far enough so that the magnets do not interact, and now move your hands around and around in any direction you choose. No matter how far those hands are spread, or at what angle they are to each other, each magnet is exercising a gravitational pull. Each hand is too. But gravity is weak, as evidenced by the fact that you cannot feel the magnets or your hands pulling toward each other. Newton taught us that the gravitational attraction between two objects is proportional to their mass and inversely proportional to the square of the distance between them. So if the gravitational attraction between two 1-kilogram objects 1 meter away from each other is defined as 1G, the gravitational attraction between those same 1-kilogram objects now 2 meters apart from each other would be ¼G, 3 meters apart would yield ⅑G, and so on. Distance dilutes gravitational attraction very effectively and is a far more important component than mass. The converse of course is that when distances decrease linearly, gravitational forces increase geometrically.

As planets revolve, they go in and out of alignments that amplify, modify, and/or cancel out each other's magnetic and gravitational effects on the Sun. The Sun, to be sure, exerts its own immense influence, but as a giant incandescent blob it is also more susceptible to twists and yanks than the harder, denser planets orbiting around it.

It all kind of knocks the Sun down a peg. Ever since Copernicus burst our collective ego bubble and convinced us that the Earth revolves around the Sun and not vice versa, the Sun became next in line to the Almighty. And though it has filtered into general knowledge that the Sun is just one of a gazillion stars, part of an immense galaxy a gazillion of which form a damn near infinite universe, we don't see or feel any of that, not like the sunshine, or even the light of the Moon. So the notion that we, tiny Earth, can actually disturb the great boiling Sun remains about as sacrilegious and scary as the notion that we, tiny human beings, can actually harm God.

PLANETARY TIDAL WAVE

One would expect the Solar System's center of mass to be located somewhere within the Sun, which is far more massive than all the planets, moons, asteroids, and comets put together. In fact this center of mass is constantly shifting, due to orbit patterns and planetary alignments, and can move to a point as far as 1 million miles (1.6 million kilometers) away from the Sun, my colleague Thomas Burgess explained to me. Burgess is a solid-state quantum physicist who has divided his career between Livermore Laboratories near Berkeley, California, and Sandia National Laboratories in Albuquerque, New Mexico.

Imagine your own center of gravity no longer located within your body but rather tugged toward some exterior point. You would of course lean in that direction, and you would adjust your movements accordingly. The Sun does not lean, but rather wobbles, and also bulges in the direction of the Solar System's center of mass. The stronger the gravitational pull on the Sun, the likelier it is that the Sun's surface will fissure, suddenly releasing what is known as imprisoned radiation, a term that describes the unfathomable amount of radiation trapped inside the Sun, sometimes for tens of thousands of years. Under normal circumstances, this radiation leaks out of the Sun in a more or less steady stream, but when the Sun's surface is pulled apart, the imprisoned radiation can be released in major outbursts.

"Imprisoned radiation could escape the Sun's surface through a rip, or even a negative bulge," said Burgess, explaining that a negative bulge, or depression in the Sun's surface, would mean that there was less mass for the radiation to work its way through.

The next peak in the planetary tidal force, essentially the sum total of the planets' gravitational pull on the Sun, will come late in 2012, according to Burgess's calculations. The sunspot maximum, coincidentally also due in that year, will compound the situation, subjecting the Sun to maximum stress. The Sun's magnetic poles, which reverse every twenty-two years, at the peak of every second cycle, are also expected to switch in 2012, adding further volatility to the situation.

The resulting synergy of gravitational and electromagnetic pressure on the Sun cannot help but distort and distend its surface, releasing megabursts of imprisoned radiation, quite possibly ones that are far deadlier than any the Earth has encountered since homo sapiens has been around.

SPACE

What a party! Four astronomers, all with Ph.D.s, plus an engineer, a physical chemist, and me, a literature graduate student, all of us from the University of California at San Diego, were drawing diagrams, laughing, drinking, and arguing about natural phenomena. At around one in the morning, Ernest, the youngest, brightest astronomer, grasped his face in his hand, squeezed really hard, and then announced, "The laws of angular momentum prove that the universe is isotropic."

A hush fell over the kitchen. Everyone's mind was boggled, particularly mine, since I had no idea what the hell he was talking about, but seeing how impressed everyone else was with his observation, I wrote it in my notebook before I went to bed.

Several weeks later I was at an elegant party in the hills of La Jolla, and a physicist from San Diego State was going on and on about how physics was the deepest reality and everything else was derivative from it and therefore of secondary importance. This was 1977, and the Big Bang theory was supplanting Genesis as our foremost creation myth, and the first one based on facts. The following year, 1978, Arno Penzias and Robert Wilson would win the Nobel Prize in Physics for their discovery that some of the ambient microwave radiation in the universe is in fact left over from the primordial Big Bang explosion. So a half dozen of us, including my faculty adviser, were circled around this physics professor, who was pretty much taking personal credit for unraveling the secret of the cosmos, and I was much, much unhappy about that because my date, Priscilla, a surfer-pretty linguist, was eating up his every word. There was only one thing to do: "I have come to believe that the laws of angular momentum prove that the universe is isotropic," I calmly observed.

If anyone had asked me what I meant by that, or hell, even asked me to repeat what I said, I would have crumbled. Somehow, the way electrons spin shows that the universe is expanding equally in every direction, though how Ernest had made the leap from subatomic to damn near infinite, I had not a clue. But those words were not just words, they were an incantation. The physics professor said simply, "Why, that's a very large statement," then shrank away from the group to sit and ponder.

Fast forward twenty-eight years. My interview with Alexey Dmitriev, for which I have traveled from Los Angeles to Siberia, in the wintertime, no less, was about to be cut off after ten minutes. Much of this section is devoted to Dmitriev's iconoclastic theories about the heterogeneity of the space-time continuum, a very complicated conversation, particularly when mediated through a Russian-English interpreter. I got confused, asked a question that missed the point entirely, got flustered, and asked an even stupider question, at which point Dmitriev began checking his watch and looking for an exit. There was only one thing to do. "But I always thought that the laws of angular momentum prove that the universe is isotropic," I confided.

Compassion showed on Dmitriev's face, and he leaned across the table. "We all believed that, Larry. I even taught that to my students when I was younger. It's nothing to be embarrassed about," the scientist soothed, and then added, "Knowing what we know now, I sometimes wish it were true."

Two hours of great conversation later, plus follow-ups and related interviews, I came away with the clearest scientific indication yet of why our planet, in fact, the whole Solar System, may be headed for disaster in 2012 or thereabouts. As for how the laws of angular momentum prove that the universe is isotropic, I still don't know, and I don't want to know.

Why break the spell?

————

8

HEADING INTO THE ENERGY CLOUD

"Delta Flight 2012. Delta Flight 2012. Now boarding Zone 7."

Leaving Los Angeles for Siberia, two numbers popped up on my boarding pass: 2012 and 7, which is my birthday/lucky number. Researching 2012 was making me a bit superstitious. Do I get on this plane? Or is this a good omen for my research? And if it was a good omen for my book, was it bad news for the world? I was more than a little confused. Sunspots, hurricanes, earthquakes, volcanoes, solar physicists, Mayan shamans . . . a fresh perspective was needed. I was headed to Novosibirsk, the capital city of Siberia, to meet Dr. Alexey Dmitriev, a geophysicist with the Russian Academy of Sciences, to learn about the galactic danger zone imperiling the Sun, the Earth, and our entire Solar System. As the Sun orbits the center of the galaxy, it encounters different areas of space, some more highly energized than others. According to Dmitriev, red alert lights are flashing about the interstellar thunderstorm we're moving through now.

"Increasing solar activity is a direct result of the increasing flows of matter, energy, and information we are experiencing as we move into the interstellar energy cloud. New demands are being placed upon the Sun, and we are experiencing the impact of these demands on our own planet," Dmitriev has written. "The time until central scenarios of the global catastrophe will become a reality does not exceed two or three dozens of Earth turns around the Sun. There is no exaggeration: in fact, we believe this prediction is rather 'soft.' "

Dmitriev, age sixty, has an impressive résumé, with more than 300 academic journal publications, mostly on geophysics and meteorology, both terrestrial and that of other planets. He has written several scholarly books and has received numerous citations and awards, including the Symbol of Honor (Znak Pocheta), a Soviet prize for his achievement in developing methods for prospecting important minerals, such as nickel, iron, gold, uranium, and oil.

For all his stack of credentials, there was no guarantee this man wasn't a nut. Attempting to contact Dmitriev had been so frustrating that I came very close to canceling the whole Russian trip. He was never in his office, and the first three times I called his home, he or his wife hung up on me. English-speaking intermediaries intervened, set up telephone appointments that were broken with excuses like, "Dr. Dmitriev is out researching thunderstorms. He'll be back in a month." It took ten weeks to get him in a conversation, at which point he suggested that I fax my questions to him. This seemed constructive, so I had them translated into Russian, faxed, and e-mailed, but they were never answered.

Dmitriev's work on the interstellar energy cloud puts him in a great tradition of Russian space science. Indeed the Russians may one day be seen to have been as obsessive about studying outer space as the Maya were about studying the skies. With an economy that turned out to be about a quarter the size of our own, and living standards even lower, the Soviet Union had managed to match the United States step for step in the space race for decades. Starting with the successful launching of Sputnik, the first satellite, in 1957, the Russians achieved the first successful lunar probes, Luna 2 and Luna 3, in 1959. They put the first astronaut, Yuri Gagarin, in space in 1961 and established the first space station, Salyut, in 1971 and the first long-term functional space station, Mir, in 1986, which operated until 2001.

I was really curious to meet Dmitriev and his colleagues, but when the plane started its descent into Moscow, where I would pass a couple of days, I found myself hitting the brake, the way a nervous passenger might do in a car headed for trouble. Best I can figure, my right leg was channeling the spirit of my father, who would not have been pleased, not one bit, that his son was heading for what to him would always be the Communist capital of the world.

Dad had been a prisoner in Italy during the Second World War and accounted himself lucky to have been captured; his two best buddies died in foxholes on either side of him in battle. Also, thank God, he wasn't the same Edward D. Joseph that the War Department thought he was when they mistakenly sent a telegram to his parents saying that all his arms and legs had been amputated as a result of injuries in battle. His mother absolutely refused to believe the news, went down to St. Anthony's Maronite Catholic Church in Danbury, Connecticut, dropped to her knees at the entrance, and crawled all the way down the center aisle to the altar, crying, begging, and cursing the Lord. It worked.

Back home, after six months or so of getting up in the middle of the night, running out to the backyard, digging trenches and then jumping in, shouting "The Jerries [Germans] are coming! The Jerries are coming!" my father pretty much got his life back to normal. Politically he was an antiwar Republican—"America: love it or leave it"—though he once declared that if I ever got drafted he'd take me out back and shoot me himself.

Fascism fell so hard, so quickly, that many patriots, including my father, needed something new to fill the enemy void. That something, quite conveniently, turned out to be communism, fascism's ally back when Hitler and Stalin were still hitting it off. Two of the very few times Dad ever got angry at me had to do with communism. Like many boys growing up in the early 1960s, I wanted to be an astronaut, of which my father was proud. One day he introduced me to a man I now realize was his new boss, and asked me to tell this man who my hero was, expecting it to be John Glenn, the first American to orbit the Earth.

"Yuri Gagarin," I said brightly. The Russian cosmonaut of course was the first man in space and also the first to make an orbit.

The other time he got mad, I was coming home from school, second grade or so, and while climbing the stairs to our apartment, I for some godforsaken reason began singing, "Communist Mommy, Communist Mommy." All the ems must have sounded good together. Dad blew his stack.

It was the height of the Cold War, the tail end of the McCarthy era, when even the most baseless accusations of communist sympathies could ruin lives.

Until the Berlin Wall fell in 1989, a book with "Apocalypse" in the title would likely have concerned the impending nuclear holocaust between the USA and the USSR. In fact in 1986 I worked on an ABC miniseries, *Amerika,* a fourteen-hour postapocalyptic marathon set some indefinite time after the commies had gotten the drop on us and we surrendered to avoid a massive and futile nuclear conflict. The action of the story was the Soviets' dismantling of our infrastructure and carving our nation into separate, helpless republics, until Heartland, the republic that comprised Kansas, Nebraska, and thereabouts, heroically rebelled. Now ironically the Soviet empire has itself been carved into independent republics, many quite helpless. And at least one, Georgia, is in open rebellion.

Riding in from the airport to central Moscow, I could not help thinking that this was still enemy territory, regardless of geopolitical cant. Is the Cold War over or just on hold? Like many baby boomers, I grew up with the image of Premier Nikita Khrushchev pounding his shoe on his desk in the United Nations General Assembly, shouting "We will bury you!" to the United States of America. Surveying the dowdy, crabby crowds shuffling along the Moscow streets, I couldn't help but think, "These are the people who almost took us out?" (No doubt Russians are similarly confounded when first beholding the Big Mac/Mickey Mouse Americans.)

Walking through Red Square, where the Soviets used to parade their nuclear arsenal every year on May Day, was a good reminder that the prophecies for 2012, a U.S. presidential election year, a year when the Summer Olympics would be held in London, capital city of our fraternal ally, could also be fulfilled by man-made catastrophe. Certainly some of those nukes have made their way out of Russia and into the hands of shadowy malefactors just waiting for their chance to strike. Are fears about 2012 becoming so widespread as to become self-fulfilling? Will an enemy who wants to psych us out choose that iconic year to attack? Will this book, if successful, make the 2012 target date that much more tempting?

Back at the hotel, I checked e-mail and retrieved a message from Dmitriev that said he would "do his best" to keep our appointment.

If I travel halfway around the world to Siberia and this guy flakes . . .

The whole Russia adventure was beginning to feel like a terrible mistake. So the trip wouldn't be a total loss, I ran around Moscow soaking up as much culture as I could, when up popped an omen, in the Pushkin State Art Museum: El Greco's portrait of John the Baptist, my absolute favorite painting in the world. El Greco saw John as a sensitive pagan, almost prehuman, yet with all the depth of character he could ever need for his sacred mission of readying the world for the Son of God.

I had seen the painting only once before, June 22, 2000, the last time I met with James Lovelock, the protagonist of my first book. It was in Valencia, Spain, where the American Geophysical Union held the second of its weeklong Chapman conferences on Lovelock's Gaia hypothesis, which holds that the Earth is essentially a superorganism, not an inanimate chunk of rock and water. In twenty years of writing about it, I had come to fancy myself the Boswell to the Gaia movement, even though Samuel Johnson's most famous biographer was, in personal life, an indecent lout. I fell in love instantly with El Greco's touching, slightly extruded tribute to St. John, the kind of man one could aspire but never rationally hope to be. That my wife, Sherry, was born on June 24, St. John the Baptist's feast day (indeed quite a festival day in Europe), added to the feeling of special connection.

Stumbling onto that painting in Moscow, so far away from its home in Spain, gave me quite a start. (Turns out there are four virtually identical El Greco portraits of St. John the Baptist, in Valencia, Moscow, San Francisco, and one other location.) But you could have handed me my head on a platter when several days later in Siberia I finally caught up with Alexey Dmitriev. He was a dead ringer for James Lovelock, wearing a so-perfect-it-looked-fake mustache. No more than five pounds, one inch, and zero shades of hair whiteness separated the two slight, sparkling scientists. Both live and work off the beaten track (Akademgorodok, a small town in Siberia; St. Giles-on-the-Heath, a hamlet deep in the English countryside) and both bear compelling messages about the fate of the Earth.

And both, it turns out, are celebrities in their field. The man whom I feared would turn out to be some crank professor gone off his Trans-Siberian rails—he predicts, after all, that we are flying into an energy cloud that will jolt the whole Solar System up, down, and sideways—turned out to be engaging and debonair, a celebrity who had given me the runaround because, like any other celebrity, he was so much in demand. He had to be extra careful with his time.

THE SOLAR SYSTEM IS HEATING UP

"I would like to state something at the beginning of this interview. There are three important energy sources denied or completely downplayed by orthodox scientists. These are (1) the dynamic, incremental conditions of the interplanetary medium, (2) energetic effects of the planetary configuration of the Solar System, and (3) impulses from the center of the galaxy," declared Dmitriev.

These are three vast statements, all with implications for 2012.

For starters, Dmitriev believes that the entire Solar System is heating up. Think global warming, to the zillionth degree.

Most of us learned, with a shrug, that we are always moving in ways we cannot feel. Beyond the Earth's daily rotation and its yearly revolution around the Sun, we are passengers of the Solar System, which is moving on some unspecified orbit through the Milky Way galaxy, which in turn is moving God knows where through the universe. Ancient Mayan astronomers, of course, studied this intently, but to us the motions of the Solar System and the Milky Way galaxy seem irrelevant, a cosmic technicality. No one ever mentioned the possibility that the Solar System might actually move into a new and possibly hostile set of circumstances, though it stands to reason that eventually it would. Interstellar space comes with no guarantee of remaining uniformly black, cold, and void.

We all are passengers on a plane, the Solar System, and our ship is moving into some stormy weather—interstellar turbulence, to be exact.

If nothing else, Siberians know their storms. I had finally caught up with Dmitriev at the International Symposium on Heliogeophysical Factors in Human Health, November 15–16, 2005, hosted by the Scientific Center of Clinical and Experimental Medicine of the Siberian Branch of the Russian Academy of Sciences, in Akademgorodok, where Dmitriev has lived for most of his academic career.

Akademgorodok is a marooned utopia, founded in the late 1950s, 30 miles outside of Novosibirsk, to be the center for top-secret Soviet research in weapons development, space applications, experimental medicine, and parapsychological research, which was deemed an aspect of espionage and weaponry. The best and brightest of Russian science were not exactly exiled to this woodsy lakeside village, just sheltered there, away from the temptations and prying eyes of the West. There's not a hint of gulag about it.

Akademgorodok came fully equipped with better facilities, better housing, and far more intellectual and cultural freedom than one could expect out of the Soviet state. The closest thing to Woodstock in Soviet history was the annual May Day festival in Akademgorodok. Plus, there was a cool hangout café, with poetry and music and all sorts of other subversive delights.

Today, much of the secret military research conducted in Akademgorodok is being declassified. The dissidents' café is now a bank. Intel is said to be building an industrial park close by. And the New York Pizza shop has a Statue of Liberty, lit up in white neon. But the younger generation is not entirely convinced. They are leaving Akademgorodok for Moscow, for the West, though in relative terms the town isn't doing badly. Its population is dropping by only a few tenths of a percentage point each year, while official projections for the Russian nation as a whole see a loss as high as 25 percent over the next two decades, with the median age jumping a decade or more.

Dmitriev's generation, however, is dug in for good. Their salaries wouldn't get them a phone booth in Moscow, and besides, their academic tradition of sequestered experimentation, freedom to pursue whatever might give Mother Russia an edge, would be impossible to match. And there are still the Soviet holdover perks. At the conference where I met Dmitriev, I bought lunch for my interpreter, Olga Luckashenko, a brilliant young doctoral student, and myself, for about $1.50, including beverage, soup, sandwich roll, and dessert, but no napkins.

JUDGING FOLKS by their clothes is a vulgar habit. Shiny suits, stiff white shirts, and polyester ties surrounded Dmitriev, who seemed, by comparison, draped in cashmere (blend). He had just published a new scientific text on the space-time continuum, and after taking a moment from our interview to autograph some copies, he explained that the notion that interstellar space is not homogeneous makes the utmost common sense and is consonant with the kind of understanding we should be developing half a century into space exploration.

Consider the high seas. The first explorers tended to assume that the ocean was homogeneous, with pretty much the same characteristics, water and waves, everywhere. That was a good first assumption, in that it allowed navigators to proceed with confidence. Then, as more firsthand experience was gained, came greater discernment regarding wave heights, water depths, currents, sea floor, and rock and coral formations. This did not fundamentally overturn the original assumption that, for example, the ocean is everywhere made of salt water (undrinkable, as outer space is unbreathable), deep enough to drown in, treacherous or potentially so. But from the glassy tranquility of the Indian Ocean—the unlikeliest source, seemingly, for a killer tsunami—to the stormy North Atlantic, ocean voyagers came to discern vital differences in what was originally considered uniform.

The same idea holds true for outer space. It was a good first bash that it was all pretty much a vacuum, and that within the Solar System conditions vary primarily according to proximity to the Sun. Interstellar space, of which we have even less direct knowledge, was therefore assumed to be even more void of characteristics. Of course Dmitriev is by no means the first to recognize these (retrospectively) obvious facts, but he is certainly a leader in understanding how the heterogeneity of space affects our current situation.

Like a pilot barking orders to fasten your seatbelts, or a captain shouting commands to batten down the hatches, Dmitriev is telling us that the turbulence ahead is not just theoretical but a fact that must be faced immediately.

To visualize what's happening to our Solar System, forget the standard, Tinkertoy models that we have all seen hanging in a classroom or museum somewhere. Imagine instead a great sphere of light, known as the heliosphere. This sphere is brightest at the center, where the Sun is, and grows dimmer the farther out one goes. The various planets, moons, asteroids, comets, and debris are doing what they always do, spinning, orbiting, and whizzing about within this great big light ball. That heliosphere, in turn, is trucking through space, perched on an arm of our galaxy, which in turn is also spinning and flying.

For a long time we just assumed it would always be smooth sailing. Now, Dmitriev explains, the heliosphere has hit a rough patch, specifically, magnetized strips and striations containing hydrogen, helium, hydroxyl (a hydrogen atom joined by a single bond to an oxygen atom), and other elements, combinations, and compounds: space debris, perhaps the result of an exploded star.

Like any other object traveling through any other medium, a boat pushing through water, for example, the heliosphere has created a shock wave out in front of it as it pushes away particles of interstellar space. That shock wave has become larger and thicker as the heliosphere has entered this denser region of space, where there are more particles to push out of the way. Dmitriev estimates that the heliosphere's shock wave has expanded tenfold, from 3 or 4 AU, to 40 AU or more. (The unit AU, or astronomical unit, is the distance from the Earth to the Sun, approximately 93 million miles, or 150 million kilometers.)

"This shock wave thickening has caused the formation of a collusive plasma in a parietal layer, which has led to a plasma overdraft around the Solar System, and then to its breakthrough into interplanetary domains . . . This breakthrough constitutes a kind of matter and energy donation made by interplanetary space to our Solar System," writes Dmitriev, in his controversial monograph "Planetophysical State of the Earth and Life."

In other words, the shock wave has wrapped around the leading edge of the heliosphere, the way flames wrap around the front and the sides of the Space Shuttle as it reenters the atmosphere, except that the Space Shuttle has specially designed shields to protect it from being torched. According to Dmitriev, the shock wave is now pushing into our heliosphere, penetrating into regions where the heat shields, had the good Lord seen fit to equip our Solar System with some, would have been placed. The net result is that large amounts of energy are being injected into the interplanetary domain, jolting the Sun into erratic behavior, distressing the Earth's magnetic field, and quite possibly exacerbating the global warming our planet is experiencing.

Dmitriev and his colleagues discovered the shock wave by analyzing Voyager satellite data from the outer reaches of the Solar System. The mission comprised two satellites and launched in 1977, taking advantage of a rare alignment of Jupiter, Saturn, Uranus, and Neptune, such that the planets' gravitational fields could be used to accelerate the satellites through space at speeds otherwise unthinkable. Voyager I and II transmitted detailed information about the moons, rings, and magnetic environments of the outer planets for over a decade, then in 1988 headed out for the heliopause, the boundary between the Solar System and interstellar space some 10 billion miles (16 billion kilometers) from the Sun.

Using the Voyager data as a baseline, Dmitriev and his colleagues compared it with more recent research culled from Russian and Western scientific journals, as well as NASA and ESA data. They found startling, consistent evidence that, from the tiniest frigid moons circling the outer planets to the heart of the Sun itself, the heliosphere is behaving in a more excited and turbulent manner than it did twenty years ago, when Voyager took its first measurements.

The interstellar energy cloud has been much studied by Russian scientists, notably Vladimir B. Baranov, who in 1995 was named Soros Professor at Moscow State University. This honor is conferred by George Soros, the peripatetic billionaire philosopher renowned for "collecting" creative and scientific geniuses. Baranov's work on the hydrodynamics of interplanetary plasma and the deceleration of the solar wind by interstellar medium has been published widely in Russian, including in the *Soros Educational Journal*. Baranov has developed a mathematical model of the heliosphere based on Voyager data. At a 1999 Moscow conference honoring his sixty-fifth birthday, planetary scientists from Russia, Europe, and the United States examined Baranov's model, which indicates a 96 percent correspondence between Voyager data, more recent NASA and ESA information, and the basic energy and space assessments made by Dmitriev, who expects our heliosphere to remain in the shock wave for the next 3,000 years.

The shock wave is strongest at the leading edge of the heliosphere as it moves through interstellar space, just as the wake of a boat is choppiest in front, at the point where the hull first slices through the water. Thus the shock wave most severely impacts the atmospheres, climates, and magnetic fields of the outer planets: Jupiter, Saturn, Uranus, Neptune, Pluto, and now the newly discovered tenth one, Planet X. (Astronomers now question whether Pluto and Planet X truly qualify as planets, but except for a sentimental reluctance to demote Pluto, that debate is beyond the scope of this book.)

Uranus and Neptune have both seen their magnetic poles shift, much the way that a number of scientists think has begun to happen on Earth. And both planets' atmospheres are shining more brightly and appear to be warming up, which is what would happen in the event of fresh energy inputs. Auroras, the spectacular light displays caused by sudden injections of radiation into an atmosphere, have newly appeared on Saturn, which in late January 2006 treated astronomers to a Mars-sized thunderstorm, with lightning bolts a thousand times stronger than those on Earth. Yellowstone-like geysers have for the first time been seen erupting on Enceladus, Saturn's moon.

Jupiter is showing some of the most pronounced effects of the shock wave. The heliosphere's largest planet has seen its magnetic field double in size, now extending all the way to Saturn. Magnetic fields are, quite literally, fields of energy; for them to double in size requires doubling the amount of energy that sustains them. From Earth, Jupiter's magnetic field, if it were visible, would now appear larger than the Sun to the naked eye. Auroras have been sparking between Jupiter and Io, its moon, which has also shown unprecedented volcanic activity. But the shocker of all was the March 2006 discovery that Jupiter is growing a new red spot, essentially a never-ending electromagnetic storm, about as large as the Earth.

Astronomers have been tracking this new red spot, officially known as Oval BA, since the year 2000, when three smaller spots collided and merged into the new conflagration. Oval BA has grown to about half the size of Jupiter's original Great Red Spot, the most powerful storm in the Solar System, which has been raging for at least 300 years.

"We've been monitoring Jupiter for years to see if Oval BA would turn red—and it finally seems to be happening," reports Glenn Orton, an astronomer with the Jet Propulsion Laboratory (JPL), in Pasadena, California. Orton explains that Oval BA's deepening red color indicates that it is growing and intensifying as a storm. From where is the power coming to supply this storm? JPL offers no explanations. Dmitriev and Baranov point to the shock wave, which is blasting energy into Jupiter's atmosphere, spurring electrical storms and the eruption of volcanoes.

The shock wave's effects have begun to be detected in the inner planets as well. The atmosphere of Mars is becoming denser and therefore more potentially biofriendly, since a denser atmosphere provides greater protection from cosmic and solar radiation. Venus's atmosphere is changing in chemical composition and optical quality, becoming more luminous, a good indication that its energy content is increasing.

Although the Sun is at the center of the heliosphere, and therefore the farthest removed from the shock wave's effects, it is much more susceptible to energy infusions than the planets are. Just as water cannot absorb water and earth cannot absorb earth, the Sun's molten mass of energy cannot absorb and dissipate energy nearly as efficiently as the hard, cold material bodies of the planets can. Thus even the relatively small early inputs from the shock wave are already having significant impact on the Sun.

"Increasing solar activity is a direct result of the increasing flows of matter, energy, and information we are experiencing as we move into the interstellar energy cloud. New demands are being placed upon the Sun, and we are experiencing the impact of these demands on our own planet," declared Dmitriev during our interview.

Whatever disturbs the Sun disturbs us, is the message. From Dmitriev's perspective, all the planets, including the Earth, are in a double bind, getting fallout from the shock wave both directly and indirectly through the tumult it creates on the Sun.

"There are absolutely unambiguous and reliable signs of this threatening phenomenon [the shock wave], related both to the Earth and the adjacent space . . . What really matters for us is to understand and accept them and endeavor to survive," Dmitriev added.

DOUBLE BUMMER

Should the Earth become uninhabitable, humanity's last resort has always been to flee to outer space and set up shop there. The Moon, of course, because of its proximity and now the increasing likelihood that there are substantial amounts of ice to be thawed into water and electrolyzed into breathable oxygen, has been the obvious choice. Mars and also Io, one of Jupiter's moons, have been mentioned as possibly being habitable. Indeed, something of a movement has sprung up surrounding the belief that establishing space colonies is a pressing need, rather than just the expression of our spirit of adventure:

> The mission . . . is to protect the human species and its civilization from destruction that could result from a global catastrophic event, including nuclear war, acts of terrorism, plague and asteroid collisions. To fulfill its mission, ARC [Alliance to Rescue Civilization] is dedicated to creating continuously staffed facilities on the Moon and other locations away from Earth. These facilities will preserve backups of scientific and cultural achievements, and of the species important to our civilization. In the event of a global catastrophe, the ARC facilities will be prepared to reintroduce lost technology, art, history, crops, livestock, and, if necessary, even human beings to the Earth.

So writes Steven M. Wolfe, one of ARC's principals.

Spread the risk and breathe a little easier is the idea. But if Dmitriev is right about us flying into the interstellar energy cloud, the whole Solar System will be in the same predicament as Earth, meaning that our eggs are going to have to travel an awfully long way to find a safe second basket. At the very least, we will have to find another star system to escape to, one close enough for human beings, perhaps cryogenically frozen, to survive the journey. The current front-runner in the survival sweepstakes is the Alpha Centauri system of three stars: Alpha Centauri A, Alpha Centauri B, and Proxima Centauri. Alpha Centauri A is similar to the Sun and may possibly have habitable planets orbiting it. The good news is that Alpha Centauri is also the system closest to us. The bad news is that close means 25 trillion miles away. It would take over four years traveling at the speed of light, 186,000 miles per second, to get there. Current manned space technology would get us there in, oh, half a million years or so, which, to look on the bright side, is way less than it would take for a new and wiser species to evolve if Dmitriev's energy cloud fries us down to microbes and roaches and such.

Work is under way. In southwestern New Mexico, not far from where Richard Branson, Paul Allen, and company are establishing their spaceport, there is—quite coincidentally—the mandatory neofascist/Freemason cabal secretly run out of the Vatican by rogue elements of the CIA, working day and night to liberate trillions in mob accounts (illegally and immorally held hostage by greedy international bankers), with which they will buy up tracts of land, where they will create an underground city (because if the city were above ground people might start asking questions), breed special livestock and foodstuffs, and assemble a modular spaceship enabling a pod of 160 (the ideal number) or 144,000 (the other ideal number) chosen individuals to flee the Earth just before it blows apart in December 2012 and soar, using a miniature controlled nuclear fusion reactor, to a nearby star system that we, for the purposes of this off-the-record discussion, will call Rom, where a type M, Earthlike planet awaits colonization.

I have met some of these guys, and they would give me first-class accommodations to Rom, probably let me pilot the damn spaceship, if, say, I bought them a Toyota.

All things considered, I'd rather hang out with them than with the Voluntary Human Extinction Movement, whose motto, "Live Long and Die Out," does make a nice tattoo.

9

THROUGH THE
THINKING GLASS

It certainly wasn't the Georgian food, a felicitous blend of hearty Russian and healthy Middle Eastern, that gave me the nightmare. That Commie cuisine was so delicious that even Dad would have tucked right into it, though not before duly reminding me that Josef Stalin was from Georgia and was in some ways worse than Hitler. Maybe it was the ride back from the restaurant. You'd think Siberians would know how to drive in the snow. Had me praying and bowing like Rasta Cabbie. Of course bad dreams do go with the Apocalypse 2012 territory, though for the most part I'd slept surprisingly well, save for the occasional great white shark chomp on my psyche. Best I can figure, it was the Novosibirsk hotel room that got to me. The toilet had a little strip that said "disinfected." But the shower curtain definitely did not.

That the scum on a shower curtain should represent the greatest triumph in the history of the Earth, perhaps the whole Solar System, is a fact well suited to the subconscious realm of dreams. It was literally a revolt from the underworld 2.5 billion years ago by scum known as cyanobacteria that created the deadliest chaos and yet also life as we know it today.

Before the rise of cyanobacteria, life on Earth was limited to anaerobes (organisms that avoid oxygen). These bacteria hid from the sunlight, and by the process of fermentation they slowly decayed the bonds of gases such as hydrogen sulfide, plenty of which had been belched out by volcanoes. One compound the anaerobes could never break apart was hydrogen hydroxide, water, the most abundant substance on the face of the Earth. So after a billion years of gnawing in the darkness, making their prehistoric versions of wine, cheese, and soy sauce, the fermenters were replaced by cyanobacteria, who stepped into sunlight and used those powerful rays to split apart the bonds of the H_2O molecule. The hydrogen that was released combined with carbon dioxide in the atmosphere to form sugars and carbohydrates that, a little further down the evolutionary road, were metabolized by rudimentary plants.

At first the oxygen released by the cyanobacteria was absorbed as rust by metals and also by other gases, notably methane. But after a million years or so, oxygen began to fill the air, rising from less than 1 percent of the atmosphere before the cyanobacteria emerged, to somewhere near our current level of 21 percent, said Heinrich Holland, a Harvard geochemist, in our interview. All but the handful of anaerobes that managed to take refuge in the ooze or under rocks were wiped out by the oxygen gas.

This was the greatest pollution crisis in our planet's history, and it forced the cyanobacteria to pull the second trick out of their evolutionary bag. After countless failed experiments dealing with oxygen, they eventually learned how to breathe the waste product of their own metabolism. (This is quite a trick, since, by definition, waste products are poisonous to the organisms that produce them.) The effort was worth it: Aerobic respiration turned out to be marvelously efficient, some eighteen times more powerful than its anaerobic counterpart.

"Cyanobacteria now had both photosynthesis which generated oxygen and respiration which consumed it. They had found their place in the sun," write Lynn Margulis and Dorion Sagan.

Though still present in original form as, among other things, the scum on the shower curtain in my hotel room, cyanobacteria have for the most part long since evolved into chloroplasts, the photosynthesizing machines in plant cells. Photosynthesis, advanced though it is, still cannot process the Earth's unfathomably vast salt water supply. The sodium chloride and other salts in sea water electrostatically collapse cellular membranes and short-circuit the chemistry. However, according to Margulis, there are microbial communities made up of many different species that have learned how to cooperate with each other and safely remove salt from sea water, sequestering and then varnishing it so it does not go back into solution.

Imagine a microbial community that could not only desalinate sea water but then could use the fresh water left over for photosynthesis. The idea is not so far-fetched, given the now-famous story of the natural nuclear reactors at Oklo, a French uranium mine in Gabon, Africa. When the first ore shipments were readied for processing, they were found to have been stripped of the fissionable isotope U235. Terrorists were suspected. Officials braced for the worst. But everyone relaxed when it was revealed that the theft had taken place some 2 to 2.5 billion years earlier, at the time when oxygen first permeated the African atmosphere. It seems that the hitherto insoluble uranium ore was oxidized, dissolved into the ground water, and then ran into streams. There bacteria learned how to collect and process it, eventually accumulating enough U235 that a critical mass was reached. A chain reaction was started and sustained at the kilowatt level for millions of years, with the bacteria distributing the wastes harmlessly as stable fission products throughout the environment.

So if bacteria can learn how to construct and operate a nuclear reactor, they can also learn how to desalinate sea water and then photosynthesize what's left over. Firestorms would rage out of control as the waters of the oceans were shredded into free oxygen and hydrogen billowing into the atmosphere. It would be the holocaust to end all holocausts, from a human point of view. The Earth would most likely recover and adjust, just as it did when oxygen first poisoned the air, causing a new and smarter species (of bacteria) to evolve. (Fires weren't a problem during the first oxygen crisis 2.5 billion years ago, because there was basically nothing—no plants, no other organic material—to burn.)

Bacteria setting the oceans on fire is, in a nutshell, what my nightmare was about, except the whole Solar System ended up getting torched, as though the Sun went hypercritical and lost control of its thermonuclear fusion, exploding the vast black ocean of space into million-mile-high tsunami walls of fire, crashing over planet-sized islands and burning them out like crumpled newspaper.

PISSING GAIA OFF

No one could possibly say for sure if the triumph of primordial cyanobacteria was simply a step in the Earth's internal evolutionary process or if it was spurred by some extraterrestrial factor, such as the passage of the Solar System into an interstellar energy cloud. If Dmitriev is correct in believing that such clouds exist, and it seems logical that they do, then there's no telling how often we have passed through them before, or whether or not it's a cyclical thing.

"Effects here on Earth [from passing into the interstellar energy cloud] are to be found in the acceleration of the magnetic pole shift, in the vertical and horizontal ozone content distribution, and in the increased frequency and magnitude of significant catastrophic climatic events," writes Dmitriev. "The adaptive responses of the biosphere, and humanity, to these new conditions may lead to a total global revision of the range of species and life on Earth."

The concept that the Earth has a biosphere, essentially the envelope in which living things normally exist, from the bottom of the ocean to the tops of the mountains and on up into the atmosphere, is reminiscent of James Lovelock's Gaia hypothesis, which holds that the Earth behaves much like a living organism. The essence of Gaia is the negative feedback system, in which the biosphere adjusts and regulates itself to compensate for external perturbations. This process is known as homeostasis, or what Lovelock calls "the unconscious wisdom of the body." If the biosphere were suddenly to warm up, for example, as the result of moving into an interstellar energy cloud, it would find a way to cool itself down, not by consciously deciding to do so but rather in the unconscious, automatic manner that your body does by sweating. The biosphere's adaptive mechanism to increasing heat could run anywhere from increasing protective cloud cover to shade the Earth from a too-powerful Sun to detonating a supervolcano, like the one at Lake Toba 74,000 years ago that plunged the planet into an Ice Age.

There are limits, of course, to the biosphere's ability to adjust itself in order to maintain the comfortable status quo. And that ability to compensate decreases as key components of the biosphere, what Lovelock refers to as vital organs, are disabled.

Much has been written about the dangerous ecological ramifications of destroying the tropical rain forest, which acts as a giant air-conditioning system in the hottest areas of the Earth. The rain forest produces clouds, which shade the Earth, and those clouds in turn produce vast amounts of rainfall, which cools equatorial regions and helps keep them from turning into desert. Although rain forests are sometimes likened to the lungs of the Earth, they are really more like skin in their ability to produce moisture, aka sweat, and keep the planet cool. Skin is the largest organ in the human body; burn too much of it off, and there is no way to survive.

"Our favorite doom scenario is that there is a threshold rise of global temperature beyond which further rapid heating is irreversible and quite beyond our control. The threshold is somewhere between two and three degrees Celsius, and if we do nothing it will be reached in twenty to forty years. It is a problem mainly for our children and grandchildren, but worrying nonetheless," said Lovelock.

Drawing on his background in mining, Dmitriev has identified a new "vital organ" of the biosphere: "Since the Earth is a large, very highly organized organism, each of its structural units or territories, such as mountain systems, rivers, tectonic faults, ore deposits, oil fields, etc., plays a certain functional role in its life, and in its connections with the outer world. For example, iron ore deposits support the climate stability because they perform the connection between the electrical activity in the atmosphere and the electrical activity beneath the Earth's surface."

That the ability of iron ore and other metals to conduct heat and electricity might prove vital to the global ecology is another one of those palm-slap-to-the-forehead observations that seem obvious in retrospect. What more natural way of removing excess energy from the atmosphere and the surface of the Earth than conducting it down through the crust and into the mantle of the planet? Whether or not the vast metallic ore deposits were put there for that purpose, or simply perform that function by accident, is academic. The more pressing fact is that these metals have been mined aggressively since the Industrial Revolution began 150 years ago, which is also about the time that the current phase of global warming began. Dmitriev argues that removing these metals has diminished the Earth's ability to suck excess energy out of its atmosphere. This could certainly help explain some of the increasing severity of the storms, as they are supercharged by energy from the shock wave.

If the Earth is losing its ability to safely absorb excess atmospheric electricity, and we're getting more of that energy because of the interstellar cloud the heliosphere is moving through now, then there will simply be more left over for us surface dwellers to deal with. We'll have to figure out how, because we're certainly not going to stop mining for iron and other valuable metals. Are there zones of conductivity that, like the rain forests, should be specially protected for the good of the global ecology? At what point does the conductivity loss become irretrievable?

Imagine, for a moment, how unpopular Dmitriev's implied suggestion to reduce or eliminate mining might be in Siberia, which is one vast natural resource region waiting to be tapped. At 4 million square miles, larger than the continental United States, and with only about a twelfth the population, at 25 million inhabitants, Siberia begs for development. I could see the Chinese making a move against the teetering Russians. Severely overpopulated, China could happily relocate 50 million to 100 million people there within a decade. China will shortly become the world's second largest economy, with all the resources necessary to extract Siberia's minerals and oil.

Save Siberia? Dmitriev's notion is not so far-fetched. If the Amazon rain forest is the cooling system of the global climate, perhaps Siberia and other regions with vast metallic ore deposits are the (electrical) shock absorbers— an ecological service we certainly do not want to dispense with if indeed we are heading into an interstellar energy cloud injecting bolts and waves of heat, light, and electromagnetic radiation into our climate system.

GLIMPSES OF THE NOOSPHERE

If nothing else, the existence of the biosphere means that the Earth would burn very differently than the other planets in the event of my nightmare Solar System fire incinerating the interplanetary space. Life can always be counted on to put up a fight when threatened with demise. It burrows underground, goes into spore state, or sends battalions of firefighters into burning buildings, in the case of the homo sapiens species.

Not so the inanimate substances which, as far as we can tell, make up all the other planets, asteroids, and of course the Sun. Those lifeless compounds may be harder or easier to burn, depending on their makeup, but they don't fight back the way life does.

In fact the biosphere is specifically constructed to transform and diffuse heat, light, and radiation, according to V. I. Vernadsky, the Russian planetary ecologist in whose tradition Dmitriev follows. Vernadsky understood the biosphere to be the Earth's intermediary layer for dealing with the Sun and in fact with all incoming cosmic energy. Thus any changes in the Sun or the cosmos directly translate into changes in Earthly life. As he writes:

> The biosphere is at least as much a creation of the sun as a result of terrestrial processes. Ancient religious intuitions that considered terrestrial creatures, especially man, to be children of the sun were far nearer the truth than is thought by those who see earthly beings simply as ephemeral creations arising from blind and accidental interplay of matter and forces. Creatures on Earth are the fruit of extended, complex processes, and are an essential part of a harmonious cosmic mechanism, in which it is known that fixed laws apply and chance does not exist.

Ignorance of Vernadsky's towering genius is one of the most shameful intellectual scandals in the annals of American science. That's the message from the twelve-nation team of scholars who compiled *The Biosphere: Complete Annotated Edition*. This, the first English translation of any of Vernadsky's major works, appeared some seventy years after the book was published in Russian, French, and other European languages. Vernadsky is as familiar to educated Russians as Einstein, Darwin, and Mendel. Europeans are aware of his work as well. But English-speaking scientists remain woefully ignorant. Even Lovelock, the twentieth century's other great exponent of the biosphere concept, discovered Vernadsky's work only after the belated translation appeared.

It is interesting to note that the august team of Western scientists who so enthusiastically revived Vernadsky's work on the biosphere have not done the same for his writings on the noosphere. Vernadsky saw the noosphere as the mental layer that encases the planet, the sum total of all our thoughts and memories as they continue today. Developed with Pierre Teilhard de Chardin, the French paleontologist and philosopher of cosmic consciousness, the noosphere is considered the product, the wellspring, or both, of all the minds on the planet. Psychic communication can be understood as navigating the noosphere. Those tempted to dismiss the whole notion as out-landish might consider that the World Wide Web, a preposterous proposition even half a century ago, indeed embodies many of the global mind characteristics that Vernadsky foresaw.

I have glimpsed the noosphere twice in my fifty-two years. The first time I was twenty, and I was taking a four-day meditation course my mother had given me as a present for graduating college. In retrospect I appreciate that the course was a bit "meditation light," a quick survey of relaxation, mental imaging, biofeedback, and free association. Much of the last day was spent teaming up with partners and "reading cases," wherein one partner would say the first name and then mentally picture a person they knew who was ill. The other partner would use meditation techniques to access that mental picture, determine the malady, and then send the sufferer healing white light.

My partner, a kindly middle-aged woman, went first, and told me that her sufferer's name was Helen. I "went down to my level," course jargon for breathing deeply and counting backward, got a vague image of Helen, and then saw a rose emerge from her buttock. A peculiar malady. Fortunately our instructor had warned us that, once down at level, we might see things symbolically, so I deduced that something had been removed from Helen's personal area and correctly guessed that it was a hysterectomy. I accounted my effort successful but hardly conclusive, since at that time, 1974, the medical establishment was convincing women that they needed treatments now known to be optional or even harmful, and was harvesting uteri like tonsils.

It was my turn to present a case, so of course, being basically a teenager, I decided to screw around. The name of my sufferer was Dana, completely gender neutral and therefore no hint at all. And Dana was hardly a sufferer, six foot four, muscular, booming, a high school varsity basketball player who had wrecked up his knee so badly that he had to wear a complicated brace. My partner picked up on "stiffness in the leg," impressive enough, though still it could have been that I had tipped her off by unwittingly moving or otherwise drawing attention to my own leg. Then my partner added that Dana had "something in his shoulder, a piece of fruit." Well, that clinched it for me. My mother had wasted her money and I my time. Dana Burke, now an esteemed physician, has never carried fruit in his shoulder.

Just before the class ended, my subconscious finally penetrated my thick skull. I remembered that Dana had wrecked up his shoulder in another athletic injury bad enough or good enough to keep him from getting drafted for the war in Vietnam. His shoulder had a big scar on the outside and a metal screw on the inside.

Now it is possible that I somehow subconsciously recalled that screw, and through unintentional body language drew attention to my shoulder in such a manner that my partner would intuit that there was some foreign matter inside said shoulder, but that all seems a bit too convoluted. The simplest explanation being the best, my partner somehow saw into Dana's shoulder. But where, exactly, would such an image exist? Vernadsky's response would be, in the noosphere.

My second such excursion into that realm came back to me the day Olga, the interpreter, and I crunched through a snowy birch woods in Akademgorodok to a picturesque Russian Orthodox chapel, the first religious structure built in Russia during Gorbachev's perestroika (reconstruction). Inside we lit candles and said prayers for the departed. Touching as the moment was, I couldn't help thinking that prayers really don't go anywhere, which they don't, because the dead aren't anywhere, except dead. Sometimes 2012 slams one, and the image of many, many dead, with futile prayers being said over their nonexistent souls . . . Olga was still praying, so, afraid of what sarcasm might pop out of my mouth, I drifted outside the church.

The Lord may move in mysterious ways, but this was not one of those moments. Anne Stander, a hemisphere away in South Africa, did not in any way communicate with me. I simply note that, while waiting outside the church, I thought back to Johannesburg and the astrological reading she had given me about my divorce and other matters. Quite perceptive, I had thought, until she insisted that I had injured my hand five years earlier. Didn't ring a bell. Too much time at the computer had occasionally given me some carpal tunnel syndrome, but that wasn't it. Anne absolutely insisted on a serious injury to my hand five years earlier. When I got back home to Los Angeles, I had asked my wife if she could recall my ever having hurt my hand. She looked at me as though I were insane. Almost exactly five years earlier, I had cut my hand open while slicing pumpkins to make some pie. There was so much blood in the bathroom, it looked as if I had sacrificed a gazelle. Ended up getting five stitches in the emergency room.

For whatever combination of hysterical amnesia and tough-guy denial, I had completely blanked on the episode. No scars, no disability, no mention of it for years. Stander saw it clearly, but she didn't "read my mind." That information was out there somewhere—best guess, in Vernadsky's noosphere.

OLD ICELANDIC SEA CAPTAIN

Just what might happen to the noosphere in the event of the destruction of the biosphere or even the incineration of the Solar System, as per the nightmare that had me twisting in my hotel bedsheets, is beyond the speculative reach of this book. But it couldn't be good. Of more pressing concern, however, is whether or not the noosphere holds any vital information about what might be in store for us in 2012.

Sadly the one man who might best be able to see the 2012 storm coming is no longer with us. I once had the privilege of visiting with him, though. Captain Eirikur Kristofersson was 100 at the time. The short, powerfully built man with a trim silver beard had for decades commanded cutters in Iceland's coast guard. He was a true hero. The walls of his room at the assisted-living facility in Reykjavik were covered with plaques, awards, press clippings, and books chronicling an astonishing record of rescues at sea.

Set apart from all the memorabilia was an illuminated black-and-white framed photograph of a man with a dark, piercing gaze: Magnus, a friend and physician long deceased, had been Kristofersson's spirit guide since the end of World War II.

Throughout his career, Kristofersson publicly attributed his ability to "see and hear things that other people don't," as he put it, entirely to Magnus. And the sea captain practiced what he preached. In 1956 Kristofersson had just guided his vessel safely back to port during a furious North Atlantic storm, when instructions came from Magnus to put back out to sea. No radio communications, no other (conventional) transmissions were received, to which the terrified radio operator and other crewmen would later vigorously attest. Nonetheless, Kristofersson reversed course and headed back into the tempest. The ship battered its way through to a set of coordinates several hours out at sea. There they found a British ship, the *Northern Star*, which, all eight hands confirmed, had been sinking for twelve hours, and which went under right after the last crewman, the captain, was rescued. The North Atlantic near Iceland is the windiest, waviest place on Earth. No one would have survived more than a few minutes in its icy, crashing waters.

"At first I did not know where these insights were coming from, and I tried to ignore them. But once I understood that it was Magnus speaking, I had no trouble using what I learned," explained Kristofersson.

I wish I had a Magnus-type spirit guide informing me about 2012. Though it violates my sense of intellectual decorum, it also makes perverse poetic sense that the end of the world, or some awesome approximation thereof, would go undetected and/or denied by the sophisticated technological establishment, which, despite all best efforts at accuracy and objectivity, is terminally beholden to the status quo.

Once Captain Kristofersson came to terms with the source of his information, he was able to save many lives. It is our collective loss that he cannot consult his spirit guide Magnus to advise us about the 2012 storm that appears to be on its way. Should we stay away from California? Should we move there so that if Yellowstone erupts the ashes will blow the other way? Could we make it to the Moon or Mars or, if need be, out of the Solar System entirely? Or do we just sit here and take whatever comes?

A million different experts are telling us to worry about different things, and we can learn from Kristofersson's open-mindedness. He was a trained captain who used all the conventional tools and then kept an ear out for information from an unexpected place. If we can come to terms with the diversity of sources and approaches regarding 2012 as aspects of a coherent whole, then we can move forward together to prepare ourselves, our loved ones, and whatever fraction of the greater world we might be able to influence, for the coming events.

A BASEMENT IN SIBERIA

Dmitriev's admirers were closing in on us, so I made bold to ask him about 2012. He studied me for a long moment, then slid off to another subject. I persisted, but Dmitriev would not sign on to that date. But he did say, in closing, that, "The global catastrophe—hurricanes, earthquakes, volcanoes synchronizing and amplifying each other in a positive feedback loop that will spin out of control, threatening the very existence of our modern civilization—we have been talking about should probably happen in ones, not tens, of years."

After Dmitriev was finally dragged off by his gaggle, I found myself presented to Alexander V. Trofimov, general director of the International Scientific Research Institute of Cosmic Anthropoecology (ISRICA), and the head of the laboratory at the Helioclimatopathology Scientific Center of Clinical and Experimental Medicine of the Siberian branch of the Russian Academy of Sciences. He led me down to the basement and showed me a very strange machine.

The Kozyrev mirror is like a six-foot-long aluminum alloy barrel. Inside, there are a mattress and some pillows on which the subject lies. It is one of the many unusual devices invented by Nikolai A. Kozyrev, the legendary Russian physicist who was either a rogue or a genius, depending on whom you talk to and the subject covered. Kozyrev believed that time flows in beams that travel freely through the vacuum of outer space, but which are hampered by the Earth's magnetic field, much as light, which also travels freely through a vacuum, is diffused and reflected by opaque objects such as earth and clouds.

Tapping into time beams, he reasoned, would facilitate psychic communication, since much of what is considered telepathic could readily be explained by time travel—most obviously, predictions about the future.

Kozyrev and his colleagues made a study of sacred places, natural and manmade, and found that a high proportion of them happen to be in spots where magnetic field densities are low. The researchers then brought psychics to those spots, compared their performance to other locations where magnetic fields are stronger, and came up with a shelf full of data showing that the lower the ambient magnetic density the better the psychics perform.

Kozyrev therefore decided to create a low-density magnetic field of his own, a mirror that reduces the field inside of it 500 times, roughly the equivalent of what one would encounter 1,000 kilometers up in the atmosphere. Kozyrev then brought his mirrors to locations of low magnetic density—magnetic fields tend to be spottier at extreme north and south latitudes—and conducted more psychic experiments, in which an operator mentally transmits symbols to distant recipients, who draw and comment upon what their minds receive.

Apparently the results were promising enough for the Soviets to make parapsychological research an academic and military priority. Perhaps in compensation for the basic human need to worship, which of course was effectively banned, the former Soviet regime held high the banner of psychic phenomena. Telepathy was used as a weapon for everything from disabling a ballistic missile's telemetry, as portrayed by Thomas Pynchon in *V.*, to outright espionage, as was strongly rumored to have been the case when the Soviets, their infrastructure devastated after World War II and with few research and development facilities left intact, nonetheless produced a hydrogen bomb almost as quickly as the United States did. Did some unsung Soviet bureaucrat somewhere, a counterpart to Captain Kristofferson, when he was presented with atomic secrets gained through psychic espionage, overcome his skepticism, acknowledging the value of the information despite the bizarre and unscientific methods by which it was acquired?

Trofimov escorted me back to his office. On impulse, I asked him about 2012.

"Your inquiry might be better directed to our colleague, José Argüelles. You see, he's been running some experiments with us." Trofimov gestured to the Kozyrev mirror.

Was this the same José Argüelles who, through his popular and influential book *The Mayan Factor,* has done more than anyone else to alert the world about 2012? The holistic artist from Mexico who poses for publicity photos with his bamboo flute has been doing psychic experiments in a basement with Siberian scientists who wear plastic pocket protectors on their white polyester shirts? True, like the ISRICA researchers, Argüelles is an academic, having been a professor at such institutions as Princeton and the University of California at Davis. But the real meeting of minds among these strange bedfellows comes in their beliefs about space-time: "The archaeologists, of course, see the [Mayan] calendar system as just that—a way of recording time. But the question of why so much time is spent recording time remains unanswered. The suspicion dawns that the calendar is more than a calendar. Is the number system, so exquisitely proportioned, also a means for recording harmonic calibrations that relate not just to space-time positionings, but to resonant qualities of being and experience, whose nature our materialistic predisposition blinds us to?" Argüelles asks in *The Mayan Factor.*

It is precisely this ability to think outside of space-time constraints that the Siberian parapsychological researchers working with Argüelles have been studying. Space-time, or what the Maya call *najt*, is one of the most voracious quicksands in contemporary philosophy, so we will here tread lightly around the edge. String theorists have it that there are eleven dimensions, or actually ten dimensions, numbers five through ten packed tightly together, a little like netting that ensnares its four-dimensional prey, plus one extra dimension that comes into being when all the others are fully unfurled. Analogs of string theory posit that we are living in a four-dimensional "pocket" in an eleven-dimensional universe. Such disquisitions are, for the urgent purposes of this book, hopelessly abstruse, unless by chance a way can be found for us to slip out of our four-dimensional pocket and into the seven other (safe) dimensions between now and 2012.

Traditionally space-time is described in four dimensions: length, width, height, and time. That time is properly considered a dimension is, famously, Albert Einstein's observation, one initially greeted with shock and confusion but which eventually turned out to be utmost common sense. Four-dimensional space-time regards every object as an event. Take, for example, the first apartment you ever visited. Directions to it would begin with two sets of coordinates corresponding to length and width, in this case, north-south and east-west. If in Manhattan, the apartment's coordinates might be on Sixty-seventh Street between Central Park West and Columbus Avenue. Then a third coordinate, height, would correspond to whatever floor the apartment was located on. But there's one more set of coordinates so obvious that we take it for granted: the dimension of time. If the apartment building had been built in 1980, and you went to Sixty-seventh between Central Park West and Columbus in, say, 1979, that apartment would not have been there, not on any floor. Ditto if the building was destroyed before you got there. Thus the apartment is an event that began in 1980 and ended, for example, in 2012.

"The stars are the material condensation points of evolution going through stages and processes until they disintegrate or explode into supernova, and finally return to the condition of God," writes Argüelles, illustrating the "object-as-event" concept on a celestial scale.

In one sense, Einstein's observation that time is a dimension is so commonsensical that it seems more like a rediscovery of ancient wisdom than a breakthrough of vanguard science. That's certainly what the shaman Carlos

Barrios thinks. In fact, he attributes many contemporary ills to the conceptual splitting of space and time. But before we go berating Western thinking for yet another callous miscalculation, let's cut ourselves some slack by also recognizing that a fundamental characteristic of dimensions is that one can move around in them. One can travel north to Sixty-eighth Street or south to Sixty-sixth, east into Central Park or west to Broadway, and up and down the elevator to whichever floor one chooses. But one cannot travel back in time to when the building was built in 1980, or forward in time to when it finally collapses.

Time travel, particularly physical movement back and forth through time, seems every bit as impossible today as space travel did a century ago. Mental time travel, perhaps throughout Vernadsky's noosphere, is a bit more plausible. Mental time travel is what Kozyrev was talking about; while his mind may have whizzed about, his body always stayed put.

Dmitriev concurs that science must address the possibility of mental time travel, pointing to the oft-noted fact that certain animals appear to sense, hours or even days in advance, earthquakes and other catastrophes.

"Physicists cannot solve the problem of why living organisms have pre-information about catastrophic events. This forces us to change our picture of the world. The world is not simply matter and energy, but also information," he told me. An alternate explanation, of course, is that these animals are simply exquisitely sensitive to such profound events as earthquakes and pick up on, for example, the geomagnetic disturbances that precede them. By extension, predictions and perceptions made by human psychics could also be explained away simply as cases of superior sensitivity, highly developed intuitive skills that, while remarkable, do not defy the basic cause-and-effect paradigm.

Uncanny coincidence characterizes the Russian-Mayan intellectual collaboration, which dates back to World War II. There are a number of apocryphal but endearing tales about how Yuri Knorozov, a young soldier with the Red Army's invasion of Berlin, rushed into the flames of a burning library to rescue a book that, lore had it, was the only Mayan codex remaining in the world. While it turns out that there were, in fact, a number of other codices extant, Knorozov didn't know that. The young soldier accepted the challenge fate had thrust upon him, spent the next decade studying the codex, and, in the 1950s, cracked the Mayan code. He discovered that Mayan hieroglyphics, developed around 500 BCE, are particularly complicated, being partly phonetic (glyphs representing sounds in the spoken language) and partly logographic (glyphs representing whole words or concepts).

Knorozov published his results in a Russian journal of linguistics, spurring a wave of scholarly interest in Mayan culture throughout his country's academic community and eventually throughout the world. His name doesn't pop up much in Mayan circles, though. From the Guatemalan government, Knorozov got a free tour of some ancient temples, plus lunch. From the Mexican government, he received a silver medal, plus dinner.

A four-volume mathematical analysis of Mayan hieroglyphics following up on Knorozov's work was published in the early 1960s by Novosibirsk State University, setting the stage for the current collaboration between the Foundation for the Law of Time, an Oregon-based group founded and headed by José Argüelles and the ISRICA researchers in that Siberian basement.

The rules of the experiment were simple. Between May 29 and June 24, 2005, Argüelles was, at preordained time slots, to spend ten half-hour periods transmitting thoughts and images from anywhere he chose. Neither his plans for where he might conduct his transmissions nor his actual whereabouts during the transmissions were to be disclosed in advance to anyone involved with the experiment. The ISRICA team, led by Taisia Kuznetsova, a doctor of medicine in cardiology at Novosibirsk State University, was to be ready inside the Kozyrev mirror at the preset times to receive whatever transmissions they could.

In July 2006, I spent the day with Kuznetsova, a bright-eyed, professorial woman who had kept a voluminous sketch journal of the images and symbols she received during the experiment. The journal was divided into three sections because the transmissions came in three waves. Kuznetsova informed me, and Trofimov confirmed, that the first wave of information received began in late April 2005.

"But I thought Argüelles didn't begin transmitting until late May 2005," I stammered, leafing through page after page of drawings of temples, artifacts, and Mayan-looking hieroglyphs and symbols.

Two cats, two canaries. Trofimov and Kuznetsova had that resplendently smug look that most scientists only dream of having one day. They had succeeded in proving their stunning hypothesis: that telepathy and other psychic phenomena defy the conventions of time. By definition, seeing into the future requires traversing time as though it were a dimension. In this case, Kuznetsova had moved forward in time to receive images that Argüelles would later transmit from Mexico, including the Mayan pyramids at Chichén Itzá, the tomb of Pakal Votan at Palenque, the Museum of Anthropology in Mexico City, the Pyramid of the Sun at Teotihuacán, the Cathedral of San Presario in Puebla, the coastal resort of Veracruz, and the ancient Mayan city of Uxmal in the Yucatán. It is important to remember that the ISRICA did not learn anything of Argüelles's whereabouts until approximately four months after the experiment was completed.

Did this first set of images, coming a month before Argüelles began formally transmitting at the appointed hours, come when he was mentally planning his Mexican trip? The ISRICA scientists could not say, but they certainly intend to investigate.

The images in the second section of Kuznetsova's book came while she was inside the Kozyrev mirror and were received for the most part while Argüelles was formally transmitting at the appointed hours. These images repeated and expanded those of the first section but with far more color and detail. A rough sketch of a temple became an intricate drawing with steps, visitors, and background scenery.

And what of the book's third section? These images flooded in during September 2005, three months after Argüelles had visited his sites, though before the ISRICA researchers were told the specifics of his itinerary. This last set of images was the most highly developed artistically, with more of Kuznetsova's own personal response to the psychic input, including pages of arcane symbols that seemed to express narrative or commentary.

In all, there were forty-two transmitters and receivers participating in this experiment, including Argüelles and Kuznetsova. Trofimov and his staff compiled the data from the real-time transmissions and found an extremely high degree of correlation between images transmitted and images received; coefficients of correlation, known as "R" values, topped 0.7. This means that in at least 70 percent of the times when a receiver in Siberia recorded an image, that image corresponded to what was being sent from Mexico. Statistically, this is very strong evidence indeed. Generally, correlations of 60 percent or higher mean that an experiment is successful.

Trofimov confirms that the results of this experiment are in line with the mass of data that the ISRICA team has compiled since the early 1990s, data that rebuts the notion that psychic predictions are just good guesses or examples of heightened conventional perception.

The ISRICA experiments clearly indicate that psychic phenomena defy conventional cause-and-effect notions of time. As V. P. Kaznacheev and A. V. Trofimov write:

> Important results have been obtained while studying transpersonal relationships. In these experiments, images that are transferred are perceived *either 24 hours later or 20–24 hours before* [italics theirs] the moment of translation of a signal by random computer sampling. In other words, when an operator is still unaware of what programme he will transmit, his receiving partners have already perceived the process within 24 hours, describing or drawing a "future" image of it.

Now, the Argüelles experiments have shown that telepathic communication can start not just hours, but weeks before formal transmissions begin.

Has Argüelles somehow picked up on some deep truth about 2012? Trofimov quickly notes that nothing regarding the year 2012 was included in the experiment.

"But on a personal note, I would say that Argüelles is a man of unusual talents," said Trofimov, concluding that he would "therefore be hesitant to discount any impressions that [Argüelles] may have gathered regarding the significance of the year 2012.

EXTINCTION

"Frozen round waffles have been introduced in western Pennsylvania!" This, as I recall, was reported in the New Product Roundup section of Progressive Grocer magazine.

"Why frozen? Why round? Why western Pennsylvania?" I marveled. "Why am I losing touch with reality?"

If being boring is a sin, my 5,000-plus pages, half charts and diagrams, of syndicated marketing studies on the dollar volume and growth, factors in future growth, marketing shares and competitive situations, advertising positioning and expenditures, distribution channels and trends, and demographics and psychographics of consumer packaged goods, done for Packaged Facts Inc., a subsidiary of FIND/SVP in New York City, will sink me like a stone on Judgment Day. I am guilty of The U.S. Potato Market, a 250-page analysis featuring "Spotlight on Crinkle-Cut French Fries." Guilty of the candy, cookie, and snack cake market, three consecutive marketing studies that earned me the trade moniker Mr. Sweet Snack. Innocent, however, of The Toilet Bowl Cleaner Market: Trends and Perspectives, until proven otherwise.

At the beginning of this book I kind of told a lie about not having had any psychic visions relevant to 2012. I was embarrassed to admit that my spiritual epiphany was market-research-related. In the early 1990s I was writing a report on category management, which superficially is about how retailers array products on their shelves but deeper down about how there are way too many silly items clogging up the stores. For example, Sunscreen X now comes in twelve different SPFs, plus banana, strawberry, wildflower, and herbal guava scents, in travel-size pump spray, family-size moisturizing cream, multipak gel, all the way up to jumbo suppository for where the sun don't shine, when all you really want is some fucking Coppertone. The dirty little secret is that most of these iterations lose money. So why are they there? To keep competing brands of viable products off the shelves entirely.

Absurd proliferation is what population biologists identify as one of the precursors to extinction. It turns out that some species go extinct by dwindling slowly down to nothing, while others start with a population explosion, overrun their resources, develop warfare or disease, and then crash to levels lower than when the *explosion began. This is the time-honored rationale for much of the hunting that is done. Thinning the herd, like pruning the tree, is a healthier alternative than simply awaiting a die-off.*

The human population explosion is generally pegged to the Industrial Revolution, which began around the middle of the nineteenth century. Historians believe that up until then the planet's population was pretty steady at the 2 billion mark for a millennium or so. That figure has more than tripled in the past century and a half, to almost 6.5 billion today, and that's not all. People are living longer. In the West, where most of the heaviest consumption is done, lifespans have almost doubled since the Industrial Revolution, rising from roughly forty to roughly seventy-five years. So the net human impact on the planet has almost sextupled in what, by historical and ecological standards, is the blink of an eye.

Are we due to be culled for our own good in 2012? Pruned back to nubs so that we may one day flourish as never before? The resulting combination of catastrophe and enlightenment would certainly fulfill the essence of the Mayan prophecies.

10

OOF!

Sixty-five million years ago, a 10-kilometer-wide comet or asteroid slammed into Chicxulub, in Mexico's Yucatán peninsula, leaving a 175-kilometer crater right in the heart of what would one day become the Mayan domain. That impact, according to Luis Alvarez, the renowned Berkeley physicist and Nobel laureate, is what led to the extinction of the dinosaurs and about 70 percent of all other species on the planet.

No folklore tradition, including the Mayan, goes back 65 thousand years, much less 65 million. Still, one can't help but wonder if there isn't some kind of evolutionarily transmitted sense memory or something that predisposes the Maya to cataclysmic prophecies, perhaps makes them more sensitive to the cycles that underlie them. If the Big Bang is still vibrating throughout the universe 15 billion years later, as Bell Laboratories' Arno Penzias and Robert Wilson demonstrated in sharing the physics Nobel, then why wouldn't the Chicxulub impact, 200 times more recent, still resonate locally? This would certainly suggest an explanation for the Mayan obsession with the sky.

One night after returning from Guatemala, a lightbulb went on, or rather, off: When it came to stargazing, I was way out of practice. Here I was predicting the end of the world based on research on sunspots, planetary configurations, and the interstellar energy cloud, yet I couldn't remember the last time I had really looked up into a beautiful starry sky. Most of us are in the same plight. For the first time in human history, more people live in cities than outside them. Light pollution has denatured the majestic firmament into a few twinkly holes that, as we've all been taught, are inconceivably distant and therefore physically irrelevant to daily life. Civilization has been cut off from the night sky.

It was time to escape the city lights.

Every July and August, the Earth passes through the tail of the Swift-Tuttle comet, and its dust particles hit the atmosphere at 132,000 miles per hour, creating the Perseid meteor shower. Peak night usually comes on August 12, so after midnight I drove the two hours to Edwards Air Force Base, where space shuttle *Discovery*, its own kind of shooting star, had landed the day before. Parked on a dark vacant lot down the road from a residential subdivision, I got out of the car and looked up at the luscious night sky. The meteors zipped by every few seconds, leaving glowing trails that soon flickered out. But what if one of those trails didn't flicker out and instead got larger and closer, becoming "a bearded star," as Nostradamus once described it, red and fiery, changing and unstable, twisting like a burning coil?

That's what dinosaurs might have seen 65 million years ago, as the killer asteroid approached, assaulting their eyes and then also their ears with rumbling, whining, splitting sounds. Even their pea-brains would have grasped the fact that terror was on the way. Their seas were about to boil, their forests burn, their mountains melt, their lands flood, their air go putrid with stench. Did the holocaust set the dinosaurs to fighting among themselves? That's no doubt what would happen among humans, seeking refuge in the familiar, no matter how bloody and horrible.

On the far side of my car, a dozen yards away, a jackrabbit took in the show. All of a sudden it bolted. A split-second later, a huge nighthawk swooped down on the spot where the animal had stood. The bird had almost caught its stargazing prey off guard. After checking around to make sure that nothing was eyeing me similarly, I made peace with the situation by noting that at least the little animal had kept its head up. Had it been looking distractedly down, that rabbit would have been a midnight snack.

Driving home, I flicked on the cabin light to check a map. Turning a light on at night makes it easier to see close but harder to see far away. The same advantage/disadvantage logically holds true for the light of insight— new things are illuminated, but some of what one formerly might have discerned is obscured. What perspective had I gained, I wondered, and what had I lost, shining my light into the darkness of 2012?

SCIENCE FOR THE AGES

Shampoo, a sexy, funny movie starring a young Warren Beatty and Julie Christie, had a moment that some critics didn't like. Out of the blue, a character's son died in a car accident. Everyone reacted, then got on with their day. The plot of the movie didn't change as a result of the death, which prompted the criticism. But Pauline Kael, film critic for *The New Yorker,* got it right. She understood that accidents happen. People up and die. Life moves on.

I also appreciated the way they handled that moment in the film because that's what happened with my father—he had an accident and died. I suppose that should have traumatized me, but as best I can read myself, I kind of prefer my dangers to come out of the blue, the way meteorites fall out of the sky. That way I don't have to waste valuable time worrying, discussing, preparing, avoiding. One of the annoyances, frankly, of researching 2012 is that one is always searching for reasons why the year is or is not apocalyptic, constructing scenarios, evaluating evidence, contemplating contingencies for survival. I tell you, if I didn't have kids, I wouldn't have written this book, certainly not in this way. When you have children, you've got to care about the future, got to do whatever you can to keep them safe. There's no choice. But if you're single, well, "for tomorrow we die" makes for a simpler lifestyle.

James Lovelock, whom I deeply admire, wrote an article in *Science* calling for someone to write a book, let's call it *Science for the Ages*, compiling all our basic scientific knowledge. The book would be printed on stock resistant to decay and would be distributed widely, all this being done to defend against the possibility that a catastrophe wrecks our civilization, pulverizes the electronic networks, and plunges us into the Dark Ages once again. That we, the human race, might actually lose our hard-won knowledge of, say, how the circulatory system works, or how epidemics are contained, or how lasers are constructed, seems, at first thought, the jitters of a ninny. But it takes only a moment's reflection to realize that, over the course of history, flourishing times have been followed by benighted ones, when what had once been known was then lost. Look no further than ancient Greece, when so much was understood, including the roundness of the Earth. To much of the millennium and a half that followed, flat thinking of all sorts predominated.

Nothing would flatten us faster and lower than another impact of the kind that extinguished the dinosaurs 65 million years ago. After that level of cataclysm, we would need Lovelock's *Science for the Ages*, and maybe also someone to teach us how to read.

Bad News—We Are Way Past Our "Extinct by" Date

With this cheeky headline, the *Guardian* informed us that, any time now, most human beings, animals, plants, and microbes on Earth will be killed.

The newspaper was reporting on "Cycles in Fossil Diversity," an article published in 2005 in *Nature* by UC Berkeley physicist Richard Muller and his graduate student, Robert Rohde. Muller and Rohde found solid, reliable evidence that mass extinctions occur regularly, every 62 million to 65 million years. Unfortunately, the last great mass extinction, the one that took out the dinosaurs et al., occurred 65 million years ago. We are now overdue.

Muller and Rohde's mass extinction hypothesis is based on a three-year computer analysis of the 542-million-year fossil record compiled by Jack Sepkoski, the late University of Chicago paleontologist, whose posthumously published *Compendium of Fossil Marine Animal Genera* is the best reference available for the study of biodiversity and extinctions. Sepkoski spent decades in libraries digging up records of fossil discoveries. Rather than classifying the fossils by species, a term that groups creatures so genetically similar that they can interbreed, Sepkoski opted to classify them by genus, one taxonomic order above species. One example of a genus is *Felis*, which includes domestic cats, bobcats, and jaguars. The genus *Canis* includes dogs, wolves, and jackals.

Sepkoski found that the 542-million-year time period covered by his compendium divided into layers roughly 3 million years apart. He then identified the oldest and youngest layers in which each genus appeared. For example, jaguars and the other cats had not yet appeared at the time of the dinosaurs' extinction, but snakes indeed predate the dinosaurs and will presumably postdate us.

Muller and Rohde synthesized Sepkoski's mammoth compendium, analyzed the results by computer, and were shocked to find that, with clarion regularity, anywhere from 50 to 90 percent of genera (plural of *genus*) vanished every 62 million to 65 million years, the time differential attributed to the 3-million-year gap that Sepkoski had found to exist between one fossil layer and the next.

In his commentary on Muller and Rohde's mass extinction hypothesis, James Kirchner, a planetary geologist, also at Berkeley but not part of the study, declares in *Nature* that the evidence "simply jumps out of the data." I have followed Kirchner's career for seventeen years. He is a skeptical balloon-popper; he lives to poke holes and undermine questionable assumptions. He nonetheless rates Muller and Rohde's statistical evidence as more than 99 percent certain, meaning, quite literally, that the next mass extinction, on the order of the megaholocaust of 65 million years ago, is on its way.

Such an event would likely result in the deaths of billions of individuals from the force of the impact, seismic and volcanic aftereffects, and then from the breakdown in infrastructure and social order that would inevitably occur. Assuming, of course, that our planet once again manages to maintain its structural integrity after being smashed.

The epitaph of civilization will be written at Berkeley. Just as Rohde received his graduate training from Muller, Muller had earlier received his training under Luis Alvarez and was privileged to see firsthand how his now-famous professor developed the impact theory explaining the disappearance of the dinosaurs. For Alvarez, the key piece of evidence was a substance known as iridium, a dust that coats asteroids and comets. Iridium exists on Earth in only microscopic amounts, except for a liberal layer 65 million years down in the fossil record, with the heaviest concentrations at the Chicxulub crater in the Yucatán. That, along with the fact that thousands of rocks at the crater site had been smashed to bits around the same time as the iridium appeared, is the smoking gun for the impact theory.

Back when Alvarez advanced his impact theory in 1980, he suspected that mass extinctions like the one that took out the dinosaurs occurred on a regular basis; he just didn't know how often. (Remember that this is some thirty years before Muller and Rohde conducted their research.) So Alvarez challenged Muller to explain what kind of mechanism might kill off most or all of earthly life on a regular basis. Muller responded with his now-famous Nemesis hypothesis, that the Sun, like most stars of its age and ilk, has a companion, probably a barely visible star such as red or brown dwarf or possibly a black hole. Muller hypothesized that Nemesis's orbit would bring it by every X million years, gravitationally jostling the Sun and destabilizing the Solar System.

But as Muller and Rohde dug into the fossil record, they found that the mass extinctions happened every 65 million years. Why should it take so long for one star to orbit the other? Seeing each other every 65 million years is a long-distance love affair, even by interstellar standards. Might not the Sun and/or Nemesis take a lover closer to home? Muller has since backed off his theory, but Nemesis continues to attract many adherents, most of whom believe that the orbital period of the Sun's companion is much shorter, in the 26,000-year range, a more plausible pas de deux. Recall that 26,000 years is the same time it takes, looking up from the surface of the Earth, for the heavens to complete one whole rotation, for the polestar to shift from Polaris to Vega and back.

The Binary Research Institute in Newport, California, produces a stream of graphs and charts bolstering the revised Nemesis theory, purportedly showing that various wobbles in the Earth's axis and abnormalities in the Sun's orbital behavior can be attributed only to some external gravitational influence, that is, a binary companion. It's a romantic concept, in a yin/yang, positive/negative, darkness/light sort of way, and not far-fetched given that so many other stars apparently have companions. But thus far the lack of any direct observational evidence has kept the Nemesis hypothesis from gaining much traction.

So what, if not Nemesis, keeps whacking the Solar System? Planet X, considered by some to be the tenth planet, discovered in 2005 and officially known as 2003UB313. It could indeed be the "x factor." Believed to be about 18 percent larger than Pluto, Planet X currently lies about three times farther out from the Sun. However, it follows a very strange orbit that transgresses the other planets' orbital planes and occasionally brings it as close to the Sun as some of the other outer planets. Such an orbit could, theoretically, have unanticipated gravitational and electromagnetic repercussions.

Sumerian astronomers may have anticipated Planet X some 5,000 years ago, naming it Niburu. The reappearance of Niburu during the recent war in Iraq, where much of Sumeria was located, has prompted some feverish end-times speculation. But among scientists, Planet X is seen much more as a new child than as a threat to the planetary family.

Muller currently believes that every 62 million to 65 million years, the Solar System's orbit passes through a region of the Milky Way that is exceptionally gravitationally dense. He hypothesizes that the sudden, extreme gravitational tug sets off showers of comets and/or asteroids that pummel the Sun and all the planets, including the Earth. His thinking squares neatly with Dmitriev's interstellar energy cloud, which, by definition, is more gravitationally and also electromagnetically dense than the comparative vacuum the Solar System has been in. Muller's scenario also resonates with the Mayan prophecies that on 12/21/12 the Solar System will eclipse the gravitational center of the galaxy, a black hole, the densest gravitational phenomenon in the known universe, leading to apocalypse.

HEADS UP!

To picture the perils of the Earth's interplanetary environment, imagine a juggler, walking down the central aisle of a cathedral during Sunday mass, keeping three separate streams of objects moving through the air at the same time. The highest stream, almost hitting the ceiling, looks like shuttlecocks, those badminton birdie things, except each one is packed with explosives. The bad news for the juggler, in this case the Sun, is that he must really thrust each of the shuttlecock bombs almost all the way to the ceiling. The good news is that, once launched, he doesn't have to deal with it for a while— it gets way high up. Every now and then one of the shuttlecock bombs bangs into a chandelier or pillar and explodes, but for the most part no one gets hurt.

This high-arcing stream of shuttlecock bombs is analogous to the Oort cloud, at the very edge of the Solar System. Named after the Dutch astronomer Jan Hendrik Oort in 1950, and based on earlier work by the Estonian astronomer Ernst Opik, the Oort cloud is believed to contain up to 100 times the mass of Earth, spread over many millions of miles. It contains, scientists assert, millions of comet nuclei, only a fraction of a percentage of which go on to become full-blown comets every year. Comets emerging from the Oort cloud are generally categorized as long-period, meaning that they take more than 200 years to orbit the Sun. Long-period comets are generally harder to track than short-period comets and are therefore likelier to sneak up on the Earth with little notice.

The middle stream kept aloft by the juggler is also of shuttlecock bombs, which, because they go only halfway up to the ceiling, don't take as much oomph to launch but therefore come back down quicker. This middle realm is known as the Kuiper belt, which extends from Neptune past Pluto and Planet X. Kuiper belt comets are classified as short-period, with orbits of less than two hundred years, and are therefore easier for astronomers to keep track of. Many Kuiper belt comets are sucked in and destroyed by Jupiter, the largest planet with the strongest gravitational field. One example of this is the mighty Shoemaker-Levy 9 comet, which in July 1994 crashed into Jupiter, causing fireballs larger than the Earth. Had Shoemaker-Levy 9 hit our planet, life here would have been burnt to a crisp.

Comets have been associated with catastrophe since the dawn of human history. They are perhaps the most storied of celestial phenomena, figuring in religion, history, and science through the ages. Comets are believed to herald new eras, portend tragedies, and transport alien occupants. They may also be the Almighty's sperm, or some startlingly close counterpart of that, if contemporary scientific speculation is correct.

With their big white heads and long squiggly tails, comets even look like sperm cells. Over the past 5 billion years, comets have seeded Earth with vital chemicals, including certain minerals, and may even have provided our water supply, according to Louis Frank, a physicist at the University of Iowa. Frank contends that our planet is peppered on a daily basis with between 25,000 and 30,000 small, dull comets—twenty-to-forty-ton "dirty snowballs," made up mostly of ice and containing a variety of chemical impurities. He calculates that the comets deposit on the Earth's surface the equivalent of one inch of water every 10,000 years. This works out to a water layer about 7.5 miles deep since our planet was born, more than enough to account for our oceans and seas, even given generous estimates for the amounts of water bound up in organisms, or dissociated by chemical processes such as weathering and photosynthesis.

Frank's hypothesis would force us to rewrite much of Earth's history. For example, it would imply a much longer period of relatively dry existence on

Earth, and the growing amounts of water would have to be figured into the evolutionary scheme of things. NASA has given qualified support to his assertions: "NASA is not yet convinced that we know how many of these, and how much they weigh, and how much water they're providing to Earth. But it's obvious that there are dark spots in our satellite pictures, and these are incoming water-bearing objects," NASA spokesman Steve Maran told CNN in an interview.

One hint that Frank may be overestimating the amount of water being brought in by comets comes from the utterly spectacular rendezvous of NASA's space probe Deep Impact, which, on July 4, 2005, carried the United States of America's glory to even greater heights by ejecting one of its probes into the comet Tempel 1. The rendezvous occurred within one second of projected timing. Data are still being interpreted, but preliminary findings indicate that the comet is made of less ice and more dust, of a fine, talcum consistency, than previously thought.

With the image of thousands of tiny comet/sperm cells trying to penetrate the atmosphere/membrane of one big Earth/egg cell, Frank vividly portrays the Solar System as an organismic entity. Regardless of his scenario's ultimate factual accuracy, it serves to integrate our understanding of the life on our planet as a consequence of far distant processes. And as in human sexuality, every now and then an exceptionally forceful comet/sperm manages to crash through the planet's membrane and impact the surface of the Earth. These are the comets of which lore is made, and on which the scary prophecies are based. Approximately 750 comets of this potentially cataclysmic caliber have been identified and tracked, with 20 to 30 more added to the database each year.

Moment to moment, the juggler is mostly concerned with the third and closest stream of objects, less like shuttlecocks than dynamite chunks. They barely clear his head, and sometimes they break into smaller pieces and explode on the floor. This is the asteroid belt located between Mars and Jupiter. Asteroids are ignoble chunks of space debris, the probable remnants of a planet that couldn't hold its bowels together, a nameless unit whose vengeful bid for immortality haphazardly threatens its neighbors with pieces of its former self. Asteroids have neither the grandeur of comets nor their vital, seminal chemistry. In cosmic terms, they are at the bottom of the heap.

Don't worry. The poetry of the heavens, the sublime harmony of infinite wisdom, will not allow our living Earth, that most wondrous place in the Universe, to be whacked to death by space junk. Or at least that's what we should keep telling ourselves, as more and more asteroids are discovered to be whizzing nearby. None of the 200 space objects, known as bolides, expected to cross the Earth's orbit will do so within the next two centuries, but of the 1,800 or more unidentified bolides projected to be out there, no one can say for sure. Objects of 1 kilometer across or greater are estimated to hit every half million years and would likely cause global catastrophe, including the deaths of millions, perhaps billions, of people.

In March 1989, Asteroid 1989 FC, about half a kilometer wide, came within 690,000 kilometers of the Earth, crossing Earth's orbit at a place where our planet had been only six hours earlier. Asteroid 1989 FC was fifteen to twenty times the size of a bolide that in 1908 incinerated hundreds of square miles in Tonguska, Siberia, with the force of 1,000 Hiroshima atomic bombs. Luckily, few human casualties were sustained. Had the asteroid hit a more populous area, or had it plunged into the ocean, thereby creating tsunami-scale waves, the resulting devastation would likely have matched either of the century's world wars.

UMBRELLA?

There has been some debate over whether or not to develop and deploy asteroid and comet defense systems. Opponents argue that, aside from the trillion-dollar price tag, any such weapons system capable of doing the job would itself represent a greater threat to civilization than the asteroids or comet impacts it was designed to deter. Terrorists or even rogue elements in the government might possibly hijack these weapons and use them to wreak havoc on satellites and land-based targets.

Asteroid and comet defense systems have long suffered from the "giggle factor," and having former vice president Dan Quayle champion them hasn't helped any. But are the giggles really nervous titters, the whistling in the dark that comes from contemplating the unthinkably apocalyptic?

Whether or not it's a good idea to drop a trillion on an asteroid-comet defense system is beyond the scope of this book, since there is no way such a system could be up and running by 2012. It can only be observed that our planet would probably suffer more from the consequences of an extraterrestrial impact now than at most other times in history. One of the effects of global warming has been to melt glaciers that have been pressing down on tectonic plates. The seismic consequences of a major impact would therefore be markedly greater now than even a century ago, since the plates have been released to move about more freely and therefore collide more catastrophically if smashed by a comet or asteroid. Think of the difference between a boulder landing in a rock-hard, ice-covered lake versus one with a surface that had partially melted. Regardless of whether the boulder bounces off or manages to crack through the ice-covered lake, the splash would be much greater in the one with a mix of ice chunks and water.

Similarly, if volcanism is indeed a negative-feedback cooling response to global warming, then there should be a greater number of "ripening" volcanoes, such as Yellowstone and Long Valley, that will be set off by a big crash, much the way a big, old tomato would be quicker to squirt out its juice if, say, smashed with a stone than if simply left alone. All of these factors, coupled with the unprecedented density of the human population, nearing 6.5 billion, makes unfathomable megadeath, of the kind Muller and Rohde confidently predict, seem likely.

Mostly the juggler has things pretty much under control, but every so often he is jostled by someone or something, causing him to drop lots of shuttlecock bombs into the congregation. This is known as the Shiva hypothesis, advanced in 1996 by M. R. Rampino and B. M. Haggerty. Named for the Hindu god of destruction and reproduction, the Shiva hypothesis states that the Solar System bobs up and down as it orbits around the galaxy, periodically encountering gravity and energy anomalies, such as Dmitriev's interstellar energy cloud. Shiva works well with Muller and Rohde's mass extinction hypothesis, holding that the bumps in the Solar System's road destabilize the Oort cloud, setting off showers of killer comets.

The Shiva hypothesis states that, over the past 540 million years, comets from the Oort cloud have been responsible for at least five mass extinctions on Earth. We are now awaiting a sixth. The Shiva prophecies were eerily foretold by Mother Shipton, the legendary sixteenth-century seer who entered English folklore by her predictions of Henry VIII's turbulent, murderous reign, and also of the great London fire. Mother Shipton's concluding vision is of how mankind plunges into warfare and suicidal chaos as a result of the "sky dragon's" sixth visitation upon the Earth.

A fiery dragon will cross the sky
Six times before this earth shall die.
Mankind will tremble and frightened be
For the sixth heralds in this prophecy.

For seven days and seven nights
Man will watch this awesome sight.
The tides will rise beyond their ken
To bite away the shores and then
The mountains will begin to roar
And earthquakes split the plain to shore.

And flooding waters, rushing in
Will flood the lands with such a din
That mankind cowers in muddy fen
And snarls about his fellow men.

ARMAGEDDON

I was in college during the Watergate era, and most of the professors and political science students at Brown had this superior attitude that, unlike what the hysterical press was saying, presidents don't get forced out of office, at least not for the level of crime Richard Nixon was believed to have committed. Those who had professionally studied American political history, who had the experience, wisdom, and maturity to grasp the big picture, said that Nixon was not going down. But though I, crazy English major, kept losing those arguments, I knew I was right.

I hated Nixon, like a good knee-jerk liberal, but I also felt connected to him, in fact, had volunteered in the New York headquarters of his 1968 presidential campaign. Mostly it was grunt stuff, but one day they gave me a prestige job. Seems our candidate had a passion for collecting anti-Nixon memorabilia, and so for a day I got to scrounge around town looking for whatever would make Nixon furious. I scored big when I presented Ron Ziegler, who would become Nixon's press secretary, with a poster of a pretty black woman, quite pregnant, wearing the campaign button "Nixon's the One."

The night Nixon announced his resignation, I was at the ballet at Lincoln Center in New York. My mother had taken me to see Rudolf Nureyev in one of his final performances, Giselle, as I recall. They actually interrupted the ballet and rolled a television out onto the stage so we all could watch the president say farewell. Lots of people cheered. My mother couldn't abide Nixon, but she felt that it was disgraceful to cheer at such a sad moment for our nation. I cheered anyway. It felt good to be right.

Apocalypse 2012 is Watergate redux for me. All the level heads and the cool-calm establishment intellectuals are going to pooh-pooh 2012. To the extent that they are just trying to keep a panic from spreading, okay, that's fair. But I'm right this time as well. As the time draws nigh, and people begin to realize that something major, unprecedented, is going to happen in 2012, I want to be there to help people behave responsibly in the face of the threat. And do my best not to smirk.

11
LET THE END-TIMES ROLL

I always wanted to write an autobiography in which I am a minor character. I would write it that way to show how we are all role-players in the great historical saga, and how happiness comes from playing our parts. Or maybe, how we are all organisms in the great ecosystem of Life, and how happiness comes from knowing when to cooperate and when to compete. It would have been one of those complex books in which threads from my life and history were woven to make the grand scheme of things a touch more accessible.

The story was supposed to begin with a burning black boulder crashing into an Arabian desert dune, melting the sand around it into glass. Depending on who was telling the story, the mottled black boulder was either a meteorite from the asteroid belt, a fiery gift from the Archangel Gabriel, or a black, and therefore evil, chunk cast out of the bright white Moon. Three thousand years ago, or maybe four thousand, no one knows for sure, a band of Bedouins happened upon the boulder, which was now housed in a little ramshackle shrine. According to Hadith, the wisdom and folklore tradition of Islam, counterpart to Judaism's Talmud, the Bedouins were Abraham and his son Ishmael, and the shrine they discovered had been built by Adam, in a place now known as Mecca.

Abraham and his son Ishmael erected a new and sturdier shrine, called the Kaaba, to house the holy boulder. When they finished, Abraham climbed the ridge above Mecca. Four times he issued a birdcall and four times a bird came and perched upon his shoulder. At God's command, Abraham chopped up each bird and flung the pieces off the ridge. Each time, the pieces reunited, and the bird flew back to Abraham and perched again on his shoulder. Abraham descended from the ridge to the Kaaba, where he was met by Gabriel, who instructed Abraham and Ishmael in the proper way to worship his gift and glorify God Almighty. He showed them how to throw stones in each of the four directions, wash themselves ritually, wear simple white robes, and circle the Kaaba seven times—the same sacred motions faithfully undertaken by the millions of pilgrims who journey to Mecca today.

Ishmael and his mother, Hagar, were buried in Mecca, and their descendants cared for the Kaaba. But over the centuries, the shrine's original purpose of glorifying God was forgotten. A tribe called Quraysh seized Mecca, and the night after the battle, the Quraysh celebrated their acquisition, fondling the black rock and "kissing everything that was kissable," as the story goes. Word soon got out that Mecca was a good place to party, and caravans plying the frankincense trade route from what is now Salalah, in southern Oman, to Damascus, took to stopping there for some R & R.

The Quraysh began placing their idols in the Kaaba, and then someone got the bright idea of renting space in the shrine, so that pilgrims could place their idols in there too. So a millennium before Islam, and four centuries before Christianity, Mecca became a center for pilgrimage and commerce. Eventually there were 360 idols in the Kaaba, one for each day of the lunar year. The idols ranged from Allah, the magnificent one, to Hubal, the war god, to a drawing of Jesus and Mary, to Al-Lat, Al-Uzza, and Manat, the three daughters of the Moon. The greedy Quraysh even broke off pieces of the sacred black rock and sold them to the pilgrims. A stone-worshipping cult sprang up.

The Quraysh tribe gradually divided into have and have-not clans, distinguished by those who had ownership in Kaaba and those who did not. From the Hashims, one of the humbler clans, came a respected young sage named Muhammad, peace be upon him. Muhammad spoke out forcefully against pagan excesses and unholy behavior at the Kaaba. His prophecies of doom were fulfilled when Mecca, surrounded by mountains, was flooded, and the Kaaba destroyed.

All the Quraysh clans collaborated to rebuild the Kaaba, but when the work was completed, the clan leaders got into an argument about who should have the honor of replacing the sacred black boulder into its special spot in the wall. Grudgingly they decided to follow the advice of whoever happened to come along next. That turned out to be Muhammad, who solved the problem by throwing down his robe and placing the black rock upon it. He then instructed three of the clan leaders to each take a corner of the robe. He took the fourth corner, and together they lifted the rock and fit it back into its spot in the wall.

Not long after solving that dispute, Muhammad was meditating in a cave when Gabriel appeared to him, revealing the first verses of the Quran. The essence of the message was that there is but one god, Allah, and that He is merciful and all-powerful. Gabriel steadily revealed the rest of the Quran to the Prophet Muhammad, whose masterful recitation, eloquent far beyond anything ever heard in Mecca, quickly gained him a loyal following.

I am of 100 percent Middle Eastern descent, Lebanese Christian with Muslim and pagan forebears, and with a powerful Jewish affinity. I trace my lineage back 1,600 years to the Arabian peninsula to the Quraysh tribe, but not, I'm afraid, to the Hashim clan of the Prophet Muhammad. Instead my family descends from the Makhzoums, led by Abu Jahal, the greatest villain in the history of Islam. Three times my ancestor tried to kill the Prophet Muhammad, peace be upon him.

Abu Jahal was by all accounts a foul-tempered man. Everything about Muhammad infuriated him, most of all the Prophet's teaching that "There is no God but God" and that his name is Allah. Muhammad's demand that the Kaaba be cleansed of all gods and idols except for Allah was a direct threat to Abu Jahal's business interests; his clan collected rents from the Kaaba and were camel-traders to the visiting pilgrims. Abu Jahal became enraged when Muhammad further explained that those who did not accept Allah as their one true God were going to Hell.

What about all the ancestors who died before hearing of Allah, Abu Jahal demanded to know? Sadly, they are now in Hell, Muhammad replied.

Such reasoning is common throughout organized religion. Dante, a

Roman Catholic, believed that even the greatest ancient luminaries, including Virgil, who led Dante through the Inferno, were condemned because they had not accepted Christ before they died. Never mind that they died before Christ was born and thus never had a chance to embrace Him. Today the Mormons have built up the world's most impressive genealogical database for the sole purpose of going back in time to save souls that would otherwise languish in perdition.

The night after their argument about who's going to Hell, Abu Jahal grabbed a boulder, snuck up on Muhammad sleeping in his tent, raised it high over his victim's head and then . . . fled in panic. An angry, winged camel with huge snorting nostrils, sent by the Archangel Gabriel, or by Abu Jahal's conscience, depending on who's telling the story, intervened and chased him away.

The power and grace of Muhammad's faith was far beyond anything the Meccans had ever encountered. Even Abu Jahal at some level must have sensed that resistance to the rising glory of Islam was futile and pathetic. The Quran, as revealed by God to Muhammad through the Archangel Gabriel, was a sublime symphony; Abu Jahal's sputtering retorts were the random chirps of a bird. Except, that is, for the Satanic Verses, which extolled Al-Lat, Al-Uzza, and Manat, the three goddesses of the Moon.

Desert people, as the Meccans were and are, love the Moon. The Moon comes out at night, and nighttime is cool, refreshing. Unlike the Sun, which punishes those who would gaze upon it, the Moon soothes the eyes and beguiles them, changing shape every night. Moon worship does not always lend itself to moral rectitude, and over the centuries the Moon worshippers of Mecca debauched quite a bit, in tribute (yeah, right) to the three dark-eyed Moon goddesses. Muhammad was out to put a stop to the orgies. Abu Jahal did not want to see the party, or revenue stream, end. Nor was he about to trade his genuine affection for the Moon, which he could see, for Allah, whom no mortal eyes could ever behold. (It is interesting to note that the crescent Moon became the symbol of Islam.)

Satan, according to a scurrilous, ancient folktale, famously retold by Salman Rushdie, whispered verses about the Moon goddesses into the ear of the Prophet Muhammad, so that they might be included in the Quran. Muhammad refused to include them. Abu Jahal became so incensed at the exclusion of what have come to be known as the Satanic Verses that he organized a posse, one swordsman for each clan of the tribe so that no one clan would be liable for revenge. Not surprisingly, word of this assassination-by-committee plot leaked out in time for Muhammad and his kinsmen to flee from Mecca to Medina.

Muhammad regrouped, returned, and defeated the much larger Meccan force in the Battle of Bedr. Once again Muhammad was assisted by the Archangel Gabriel, who blew a sandstorm into the enemy's face. At the end of the battle, Muhammad hoisted Abu Jahal's severed head and declared, "Behold, the enemy of God!"

A fate this descendant devoutly wishes to avoid.

I cannot deny, however, an eerie resonance with my pagan ancestor. His reverence for the three Moon goddesses is mirrored in my own devotion to the Gaia hypothesis, a living Earth philosophy I have expounded since 1986. Gaia suggests reverence toward the natural world, as embodied symbolically by Mother Earth. As for the Moon, so empowering to my ancestor, I have campaigned both editorially and corporately for its colonization, in the belief that the Moon will be the Middle East, the great energy repository, of the twenty-first century. Unlike the Earth, the Moon contains vast reserves of helium-3, the ideal fuel for use in controlled nuclear fusion, perhaps the most powerful force in the universe.

That said, I personally am safely, happily monotheistic, baptized and married Roman Catholic, confirmed and comfortable with being Episcopalian. The many, many blessings of my life, particularly my two children, deserve all thanks and praise from me to Almighty God, who would not have been so good to me had He carried a 1,600-year grudge on down from my ancestor. I am, in fact, banking on His continued good will, because, as regards this book, and particularly this section, I find that I am forced to take serious exception with the literary character known as God as portrayed in certain sections of the Bible and Quran dealing with end-times.

Someone has to.

BIBLE CODE

The Bible tells us that God will annihilate the Earth in 2012.

This is the conclusion reached in *The Bible Code*, Michael Drosnin's global bestseller that plausibly purports to have decoded a secret, divine code embedded in the text of the Bible. The diamond-hard basis of this claim is a scholarly article entitled "Equidistant Letter Sequences in the Book of Genesis," written by three Israeli mathematicians, Doron Witztum, Yoav Rosenberg, and . . . Elijah, in the person of Eliyahu Rips (Eliyahu is the transliteration of the proper Hebrew spelling of the fiery prophet's name), and published in *Statistical Science*. This remarkable piece of statistical analysis verifies an observation first made by a rabbi in Prague, H. M. D. Weissmandel, that "if he skipped fifty letters, and then another fifty, and then another fifty, the word *Torah* was spelled out at the beginning of the book of Genesis." The same skip sequence yielded the word *Torah* in Exodus, Numbers, and Deuteronomy, the second, fourth, and fifth books of Moses. (For some reason, this procedure does not hold true for Leviticus, the third book of Moses, which spells out the rules for priestly comportment.)

That discovery piqued the researchers' curiosity, to see what else might be encoded. The task was daunting: Isaac Newton taught himself Hebrew and spent decades searching for the code he felt sure was embedded in the Bible. Newton, perhaps the greatest scientific mind in history, came up with nothing. That's because he didn't have a computer. The three Israeli mathematicians input the Book of Genesis, in the original Hebrew characters, letter by letter, with no spaces, no punctuation, just as biblical texts were originally scribed. In essence, they laid out Genesis as a huge acrostic and then searched for words to circle—vertically, horizontally, and diagonally. With the help of the computer, they examined this acrostic for words not just composed of adjacent letters but also of letters separated by a given number of spaces, just as Rabbi Weissmandel had originally done in finding the word *Torah*. Using the same equidistant-letter, skip-code approach, computer analysis yielded the names of sixty-six legendary rabbis, virtually all of whom lived many centuries or millennia after Genesis was written. In each case the names were close to, or intersected by, the rabbis' birth and death dates and cities of residence.

Certainly no mortal could know, and also surreptitiously encode, the names of such venerable holy men so many centuries into the future. The implication is clear: There is a secret code embedded in the Bible, placed there by God. Eliyahu Rips, Drosnin's chief collaborator, explained the seeming impossibility by citing one of the rabbis discovered in the Bible, yet another Elijah, the famed eighteenth-century sage Rabbi Eliyahu of Vilna: "The rule is that all that was, is, and will be unto the end of time is included in the Torah, from the first word to the last word. And not merely in a general sense, but as to the details of every species and each one individually, and details of details of everything that happened to him from the day of his birth until his end."

It's as though the United States Constitution were found to contain the names of sixty-six future presidents, intersected by or adjacent to their home states and the dates they were elected. Or if the 1965 annual edition of *Sports Almanac* contained the names of the next sixty-six Super Bowl winners, with the scores of the games.

The Israeli mathematicians' rigorous statistical analysis concluded that there is virtually zero probability, one in fifty thousand, of this all being a chance occurrence. Not surprisingly, their extraordinary claims came under a barrage of attacks. In the decade since this paper was published, a number of statisticians and mathematicians, including experts from the National Security Agency in the United States, have challenged their findings by disputing their methodology and by running comparable tests on two other original Hebrew texts as well as the Hebrew translation of *War and Peace*. To the best of my knowledge, no such test has been run on the Quran. But no one thus far has laid a glove on the Israeli mathematicians. Indeed, some of those who set out to refute the existence of what has come to be known as the Bible code are now among its most ardent supporters.

Drosnin, a journalist, began digging into the Bible codes for clues about the future. His most famous discovery is that the name Yitzhak Rabin is physically crossed by the phrase "assassin that will assassinate." Further deciphering indicated a place, Tel Aviv, and a date, 1995, which at that point was still in the future, of course prompting Drosnin to make every effort to warn Rabin, but to no avail. After Rabin's tragic murder, Amir, the name of the right-wing assassin, was found to be encoded nearby.

What's next? Drosnin logically wanted to know. Out poured any number of observations and predictions, mostly concerning the Middle East. Drosnin tends, as so many of us with Semitic heritage do, to equate the outcome of that region's interminable drama with the fate of the world. I have never once heard of anyone from the Southern Hemisphere imply that our collective fate hangs on the outcome of their regional disputes. Although Drosnin has been duly criticized for overinterpreting the Bible code, an impressive proportion of his predictions have since panned out, including a prediction that a judge, in this case the United States Supreme Court, would rule against Al Gore and in favor of George W. Bush in the matter of the 2000 presidential election.

Let's all pray that Drosnin's hot streak cools. Because according to *The Bible Code*, comets are expected to pound the Earth in 2010 and also 2012 (5772 in the Hebrew calendar), at which point the "Earth annihilated" prediction also comes into play. True, his analysis also unearthed the phrases "It will be crumbled, driven out, I will tear it to pieces" near the 2012 comet, though that could be a mixed blessing, causing the Earth to suffer multiple major impacts, potentially more damaging than one big blast.

The Bible Code provides the most profound scientific evidence yet that the Bible was divinely inspired. The work of the Israeli mathematicians Rips, Witztum, and Rosenberg has withstood all scientific challenges thus far. The good news is that the book upon which so much of the world's religious faith is based has received unprecedented mathematical substantiation. The bad news, of course, is how the Bible story ends.

THE ARMAGEDDON MOVEMENT

Then I saw coming from the mouth of the dragon, the mouth of the beast, and the mouth of the false prophet, three foul spirits like frogs. These spirits were devils, with the power to work miracles. They were sent out to muster all the kings of the world for the great day of battle of God the sovereign Lord. (That is the day when I come like a thief! Happy the man who stays awake and keeps on his clothes, so that he will not have to go naked and ashamed for all to see!) So they assembled the kings at the place called in Hebrew Armageddon.

Revelation 16: 13–16

Some say that from atop Armageddon, the fabled hill that looks out over the Megiddo plain in Israel, you can see the end of Time, because that's where the Battle to End All Battles will be fought. (In Hebrew, *har* means "hill," Megiddo becomes "mageddon.") Prophesied in Revelation to be the site of the final clash between Good and Evil, that is, between those who accept Jesus Christ and those who do not, Armageddon looks down on a 200-mile-long valley that will one day be filled with corpses, 2 to 3 billion of them, according to some scholars' extrapolations. It's supposed to be quite a majestic view. But you couldn't prove it by me. I will never set foot on Armageddon. And I hope you never set foot there either or, if you already have, I hope you never return.

Armageddon refers to the great, consuming war to be fought among the peoples of the Earth. Apocalypse is the natural/supernatural cataclysm expected to come after Armageddon. I am opposed to all of it, no matter how "enlightened" the aftermath will supposedly be. (If it all turns out to have been worth it, then I will crawl out of my earthen bunker and admit my mistake.) Now, trying to oppose global catastrophes such as supervolcano eruptions or comet impacts would seem about as effective as trying to oppose the law of gravity. But Armageddon is different. Of all the potential cataclysms, Armageddon is the only one for which significant numbers of Muslims, Christians, and Jews actually hope, pray, and scheme. And it's the one end-times prophecy that we actually might have the power to prevent, or fulfill.

Karl Marx observed that when a theory grips the masses, it becomes a material force; sadly, Marx's theories did just that for more than a century. The doctrine of Armageddon has gripped several small but extremely motivated and influential groups in the United States, Israel, and the Muslim world, and that doctrine is rapidly becoming a powerful, perhaps unstoppable force in global politics.

"While most Jews, most Christians, most Muslims, most everybody abhor and eschew the harness of fundamentalist thinking, history is not driven by most of us . . . As a rule, majorities are ruled. It's the fanatic few, at whom we may laugh one day and cower before the next, who are history's engine. It's a minority of single-minded maniacs who can take a holy place and make an unholy mess," observes Jeff Wells, a blogger for the webzine *Rigorous Intuition.*

Far more disturbing than its hold over a few zealots is Armageddon's powerful crossover appeal. *The Late Great Planet Earth,* by Hal Lindsey, which predicted that the great Armageddon battle would come in 1988 or thereabouts, was the best-selling nonfiction book of the 1970s. Israeli tour operators have seen their business double and double again as impassioned Christians from the United States, Europe, and elsewhere flock to the region.

Indeed, a recent survey conducted by the Israeli ministry of tourism indicates that of the nation's 2 million visitors each year, more than half are Christian, and more than half of those identify themselves as Evangelicals.

Evangelical Christians are the group most eager to precipitate Armageddon, looking forward to the Rapture, the exalted moment when, before the battle begins, true and faithful Christians are literally lifted up into the air, into the heavens, to join God. No doubt this would be exhilarating. From the safety and comfort of Heaven, one would have the opportunity to look down upon the Earth and watch the battle between two warring groups: Christians who, due to imperfections in their faith, or because of special warrior destiny, were not subsumed in the Rapture; and followers of the Antichrist, a charismatic false Messiah, whose followers include secular humanists, pagans, Hindus, and Buddhists, as well as Muslims, Jews, and insufficiently committed Christians. A large proportion of Jews are expected, in Evangelical theology, to convert to Christianity and thus fight on the righteous side of the Armageddon battle. Those who decline Jesus will, along with all other naysayers, explode.

The more people who go to Armageddon, the more mystique that hill gains, and the more likely that some incident, spontaneous or staged, will ignite a tragic war. Wave upon wave of pilgrims will soon besiege the new Christian theme park being built nearby, on a 125-acre stretch along the Sea of Galilee, where Jesus Christ is reported to have walked on water. The $50 million project is being developed by a partnership between the Israeli government and American evangelical groups. According to a spokesman for the National Association of Evangelicals, a 30-million-member group spearheading the project, the Galilee World Heritage Park should open in late 2011 or early 2012.

Things are just breaking the right way for the Armageddonists these days. What may be the oldest Christian church in the world was accidentally unearthed in Megiddo in late 2005 by Ramil Razilo, a Muslim prisoner serving a two-year term for traffic offenses. Razilo was part of a crew of inmates helping to construct a new facility to detain and interrogate Palestinians. Armageddon Church, as it is now known, dates back to the third or fourth century, a time when Christian rituals were still conducted in secret. At the center of the 24-square-foot mosaic on the floor is a circle containing two fish. The fish is an ancient Christian symbol; the spelling of the Greek word for fish makes an acrostic for the name Jesus Christ. Early Christians greeted each other by making the sign of the fish, which also alluded to the apostle Peter, a fisherman, who went on to become a "fisher of men." Peter's name, which means "rock," was an allegory for the rock upon which the Christian church was built, most famously, St. Peter's Basilica at the Vatican in Rome.

Although there is no specific biblical foretelling, this discovery is already being hailed as yet another sign that the end is near. Estimated restoration date of Armageddon Church in Megiddo: 2010–12.

DIVINE RETRIBUTION

Years ago a friend and I tried to get jobs on *Saturday Night Live* as writers. One of the skits we pitched was Yassir Arafat and Ariel Sharon doing a "Tea for Two" soft-shoe duet, dancing with blazing machine guns instead of canes. Nix.

Twenty years later, the death of Arafat, a terrorist who believed he was defending his people from apartheid and perhaps genocide, did not move me particularly. But much to my surprise, the massive stroke that struck down Sharon shortly thereafter hit me like a shovel. I had always despised Sharon for his deceitful and barbaric invasion of Lebanon in 1982, betraying Prime Minister Menachem Begin so brazenly that Begin, whose beloved wife passed away right around then, plunged into a depression from which he would never emerge.

True, Sharon, like Begin, eventually "did a Nixon" and translated his hawkish street cred into a peace process that included Israeli withdrawal from the Gaza strip. That earned respect but not affection. Of course there was sadness that hawks within Israel and without, too weak or cowardly to face down Sharon mano a mano, would now try to push the region back toward chaos. But that was only part of it. Was Sharon's massive brain hemorrhage, as televangelist Pat Robertson claimed, divine retribution? Was it God's way of saying that Israel had erred in giving up some "holy land" and/or His way of preventing the second and far more controversial stage of Sharon's plan, withdrawal from parts of the West Bank?

"He was dividing God's land," Robertson said of Sharon during his long-running *700 Club* television show. "I would say, 'Woe unto any prime minister of Israel who takes a similar course to appease the European Union, the United Nations, or the United States of America . . . God says, 'This land belongs to me, and you'd better leave it alone.' "

Robertson got spanked for his offensive remarks and was even excluded, officially at least, from the partnership developing the Galilee Christian theme park. But the televangelist really only gave voice to what many other Bible buffs quietly believe—that Sharon had betrayed the Armageddon game plan and paid the price for doing so.

SATAN IS STRONG AND GREEN

Tim LaHaye is an Evangelical preacher whose Left Behind series of apocalyptic morality thrillers have sold more than 60 million copies by demonizing the United Nations. His blockbuster archvillain is one Nicolae Carpathia, former secretary-general of the United Nations, often just referred to as "the evil one." A captivating storyteller, LaHaye has tapped into a deep well of fear among the Christian faithful that the United Nations, with its black helicopters always lurking near, is a godless enterprise out to take over the world.

Despite all the religious feuding in the Middle East, Armageddon is not a battle among Christians, Muslims, and Jews, but rather the one between those who fear God and those who do not, regardless of their choice of messiah. LaHaye, who in his spare time leads tour groups to Armageddon, may seem divisive in his pounding Christian ideology, but in fact he has helped unify the Armageddonist movement by depicting an enemy—variously described as the New World Order, the World Government movement, or just plain communism/socialism coming back in a green/pagan form—that Muslims, Christians, and Jews can together oppose.

Scuttlebutt has it that the archvillain Carpathia is modeled after Maurice Strong, the ultrawealthy Canadian industrialist, mining magnate, and environmentalist. A favorite target of Armageddonist publications such as *Endtime,* Strong is precisely the kind of shadowy, New Age figure to fit the Antichrist/Carpathia profile. He is a self-proclaimed socialist, hobnobs with world leaders by the dozen, and is a habitué of groups such as Bilderberg, the ultrasecret cabal headquartered in Leiden, Holland. Established in 1954 to be the repository of the New World Order power base, Bilderberg's members include Bill Clinton, Melinda Gates, Henry Kissinger, Tony Blair, and many others far too powerful to sport household names. Further fanning the right wing's suspicions about Strong is that his wife, Hanne, runs a spiritual, artistic, Buddhistic retreat on their vast ranch in Colorado.

Strong is considered the architect of the Kyoto Protocol, which proposed to reduce the amount of carbon dioxide in the atmosphere to a level 5.2 percent below 1990 levels by 2012. Kyoto was the culmination of a decades-long process that began in 1972 with the United Nations Conference on the Human Environment. That conference, held in Stockholm, Sweden, is credited with having injected environmentalism into the global public policy debate with a host of "green" initiatives, including a ten-year moratorium on commercial whaling. It was followed up twenty years later by the 1992 United Nations Conference on Environment and Development, the Earth Summit, in Rio de Janeiro, Brazil, which drew more than 100 heads of state and focused on saving the rain forest and other endangered ecological regions.

Strong, who dropped out of high school in Manitoba at age fourteen, was secretary-general of both global events.

Strong is an ardent proponent of world government, which of course is what any self-respecting Antichrist/Carpathia would advocate as a means to seize dominion over the whole human race. But what sets him apart, in Armageddonists' eyes, from United Nations Secretary-General Kofi Annan, to whom Strong is a close adviser, or from Ted Turner and Al Gore (both good friends), is that Strong has methodically dedicated much of his career to creating and controlling the New World Order.

Antichrist or Earth Savior? In interests of full disclosure, I worked with Strong's team at the Rio Earth Summit and would be slightly more likely to benefit from his accession to power than not, if only because those associated with the Gaia agenda, such as it is, might be more welcome than not. In fairness to the Evangelical Christians' criticisms, Strong does tend to surround himself with "godless" (not religious in any traditional sense) superstars. For example, Strong's longtime collaborator is Gro Harlem Brundtland, another off-the-radar dynamo who is one of the most beneficially influential women in the world. And a dyed-in-the-wool socialist.

A family physician who served three terms as prime minister of Norway, former head of its Labor Party, Brundtland is an ardent feminist who is devoted to her family, including her conservative columnist husband, whom she once saved from drowning. She is a former vice president of Socialist

International, a worldwide socialist network, and believes that health care is a fundamental right of humanity, necessary to the functioning of any democracy. She recently retired as director general of the World Health Organization (WHO), the Geneva-based United Nations agency responsible for increasing global health standards. Brundtland was strongly criticized for a slow response to the heavily politicized global AIDS crisis, but she was justly hailed for establishing WHO's rapid-response team that proved so effective in containing outbreaks such as Ebola and SARS.

As head of the United Nations World Commission on Environment and Development, informally known as the Brundtland Commission, she established the doctrine of "sustainable development," which regards poverty as the greatest pollutant of all. So why do we know more about Paris Hilton? Brundtland is dry, encyclopedic, has an unprepossessing appearance, the aunt who gives you a savings bond for your birthday. She is also one of the greatest healers, of people and the environment, that the world has ever seen.

With Strong, Brundtland co-organized the 1972 UN conference in Stockholm and cochaired the 1992 Rio Earth Summit. The duo are said to be planning one more UN megaconference in 2012, the goal of which is to consolidate and codify environmental precepts into binding global statutes.

"This interlocking [of the world's economy and Earth's ecology] is the new reality of the century, with profound implications for the shape of our institutions of governance, national and international. By the year 2012, these changes must be fully integrated into our economic and political life," writes Strong, who, like Carpathia, will no doubt take charge of it all.

DRAIN THE MIDDLE EAST ABSCESS

Go to the pharmacist and complain of an earache, and you'll probably get referred to the aisle where they sell eardrops. Go to the doctor, and he or she will check your throat, your sinuses, and your lymph glands, as well as your ear. The pus-filled abscess that needs to be drained from the Middle East is not, in fact, in the Middle East. It's in Europe, where the Holocaust was committed and has never been paid for.

Quick, how many Nazi war criminals were convicted altogether at the Nuremberg War Tribunals?

(a) 1,213
(b) 674
(c) 87
(d) 19

If you chose (d) 19, you are correct, and probably a little bummed.

Subsequent trials, plus the heroic efforts of Nazi hunters such as Simon Wiesenthal, and also of those who worked for the Israeli security forces, bring the total figure of disgusting murderers brought to justice or otherwise eliminated to, maybe, a couple hundred, out of a nation of 70 million that systematically exterminated up to 6 million men, women, and children, mostly Jews. Plus a few billion deutsche marks in reparations to Holocaust survivors doled out over a couple of decades. Germany and Austria got a damn good deal.

Unbelievably good, says Mahmoud Ahmadinejad, Iran's antagonistic president. He has consistently expounded a fundamental truth that, while obvious throughout the Arab world, is vehemently rejected and denied in the West. The truth: that Germany never paid for its crimes. Ahmadinejad has risen to power by giving incendiary voice to a question that has long troubled many Arab minds: If the Holocaust really happened, as the Europeans maintain, how come Germany got off so easily?

Talk about no-fault genocide. Immediately after the Second World War ended, billions in rebuilding funds from the Marshall Plan and allied sources flowed into Germany and Austria, repairing infrastructure, industry, and vital services. True, Germany was divided for a time, but that all resolved itself when the Berlin Wall came down in 1989. So either the Holocaust never happened—and every sane individual knows that it did—or someone else (the Palestinians? the Muslims?) paid for Germany's sins.

Ahmadinejad and others of his rhetorical persuasion play to the crowds by claiming that the reason Germany—and by extension the fascists in Italy, France, and Spain that allied with the Nazis—never paid for their sins is because the Holocaust never happened in the first place. This is disgusting: The Holocaust was probably the most horrible, tragic episode in human history. But one can nonetheless understand the incredulity of the average angry, propagandized Arab on the street.

Except for perhaps acknowledging a mild French affinity for Arab culture, which dates back to Napoleon Bonaparte's conquest of the Middle East at the turn of the nineteenth century, the average Arab sees the Christian powers—the United States and Western Europe, including Germany—as pretty much a unified bloc. How do you think Western preaching about the morally superior values of democracy—indeed, our highest secular value—sounds coming from the mouths of leaders whose governments, within living memory, committed the Holocaust and then basically forgave each other for it? Would you take direction from such a source?

Skeptics press the attack, saying that it's absolutely impossible that sane, civil societies, such as European Christian nations undoubtedly are, could have committed such crimes without paying for them later, thereby making things right with their victims and with God. This is when the German and Austrian government officials start getting uncomfortable. They committed the Holocaust and stuck someone else with the tab. How'd they get away with it? Because Germans and Austrians are white people who make neat stuff?

The most common rationale for this leniency is the Treaty of Versailles, which ended World War I by punishing Germany disproportionately. The Kaiser's mighty nation collapsed into the pathetic Weimar Republic, where a wheelbarrow full of marks was needed to buy a loaf of bread. Out of that chaos rose the Third Reich. But two wrongs, as they say, don't make a right: Destroying Germany and Austria after World War I does not justify a slap on the wrist for genocide during World War II.

Plato taught us to attack the strengths of our opponents' arguments, not just the weaknesses. Rhetorical attacks may help minimize hotheads like Ahmadinejad, but they do nothing to refute the truths that he, and whoever comes after him, uphold. Acknowledging the Islamic perspective on the Holocaust—that the Europeans who perpetrated it have not truly been punished and have not made amends, and that the Arab world, particularly the Palestinians, have been scapegoated as a result—in no way undermines the basic Western position that democracy must take hold in the region. We just can't be so condescending about it.

Here's a novel Middle East peace proposal. The Palestinians get the West Bank and Gaza to form an independent state. The Jews get Israel as it existed in its pre-1967 boundaries. Plus the state of Bavaria.

NOT THAT IT would be worth it, but if some great cataclysm and/or revelation does befall the world on 12/21/12, I for one would take at least a moment's consolation in the utter astonishment that the whole Bible-Quran crowd would feel at having been aced out of the most important prophecy in the history of humanity by a bunch of pagans from the boonies of Central America. Diehards would point to the Bible code's prediction that the Earth will be annihilated in 2012, but the fact of the matter is that 12/21/12 is, first and foremost, Mayan prophecy. May the good Lord protect us from any such catastrophe, but not from a near miss, not from the scare of our collective lifetimes. We could all use a wake-up call, and no one more than the hate-filled, conflict-obsessed Middle Eastern–harkening religionists who have somehow concluded that they are the ones closest to God.

12

HAIL THE STATUS QUO

My evil ancestor, Abu Jahal, had a nephew, Khalid, who ultimately rejected his uncle's pagan faith and went on to become Muhammad's greatest general. By age twenty-nine Khalid, known as the Sword of Allah, had conquered much of the Arab world in Islam's name. The Sword's most famous campaign was in 635 CE, an 800-mile dash across the desert to Damascus, the warriors stopping only to slit their camels' humps to suck out some water. At the time, Classic Mayan culture was in full flower, and the end-date of 2012 had already been prophesied by their astronomer-priests.

The Sword routed Damascus's Byzantine occupiers and entered the jeweled city of Damascus in triumph. He dismounted to honor the site where St. Paul, who had been blinded by the light of the Lord, regained his sight. On that spot the Sword knighted one of his warrior cousins, the fastest horseman in the group, Shehab, meaning "lightning." My maternal grandmother, the youngest of twenty-three children, was a Shehab on both sides.

THE ARMIES OF ISLAM took Jerusalem three years later, in 638 CE, and happened upon a place that was used as a garbage dump by the local Christian authorities. That place was known as Temple Mount, the holiest site in the Jewish religion. According to the Talmud, it was the earth of Temple Mount that God gathered to form Adam. It is also the place where Abraham proved his faith by offering to sacrifice his son, Isaac. King David erected an altar, or some say, a throne. His son, Solomon, built the first temple there, thus the name, Temple Mount, circa 950 BCE. The First Temple stood until it was destroyed by the Babylonians in 586 BCE. The Second Temple was rebuilt circa 515 BCE, then destroyed by the Roman Emperor Titus in 70 CE. The Romans, however, were unable to destroy the Second Temple's western wall, now also known as the Wailing Wall.

The Prophet Muhammad is the first Muslim said to have visited Temple Mount, in 621 CE, during his famous overnight journey. Muhammad was transported from Mecca to a spot close to the Temple Mount's western wall, and from there he journeyed to Heaven and Hell, much as Jesus was said to have done. As the story goes, Muhammad accidentally bumped a glass, and the water began to spill out. The Prophet then traveled throughout all dimensions and galaxies and back again before the water hit the table. In addition to its miraculous nature, Muhammad's overnight journey is important to Islam because it connects the religion physically to Jerusalem, and therefore to the great biblical traditions of Judaism and Christianity. In Islam, Muhammad is considered to be the final and greatest prophet in a lineage that includes Abraham and Jesus.

The Muslim conquerors cleaned up Temple Mount, then resanctified it with ritual and prayer. In 690 CE, the Dome of the Rock, a Muslim shrine, though not a mosque, was built on Temple Mount. And in 710 CE, the Al Aqsa mosque was built on the spot from which Muhammad ascended to heaven.

The Muslims lost control of Jerusalem to the Crusaders in 1099 but won it back in 1187, when the legendary Islamic warrior statesman Saladin, in whose army the Shehabs enlisted, came down from Damascus to defeat Richard the Lionheart in the Third Crusade. Jerusalem was rebuilt in the early 1500s under Suleyman the Magnificent, emperor of the Ottoman Empire; the Al Aqsa mosque has been since rebuilt and expanded a number of times and, like the Dome of the Rock, still stands today.

Control of Jerusalem's Temple Mount is one of the most potentially explosive and least discussed Armageddon issues today. Though clearly within the borders of Israel, Temple Mount remains in Muslim custody. It is controlled by the Waqf, an Islamic land trust that operates with virtual autonomy from the Israeli government. The world's defense against the catastrophe of Armageddon rests therefore on the continued close cooperation of Israeli security forces and local Muslim authorities.

Temple Mount is a holy site in Christianity primarily because Christians regard both testaments of the Bible as their heritage. But there are also a number of important references to Jesus being there. In a famous New Testament story, Jesus chased the merchants and moneylenders out of the Second Temple. After Jesus' crucifixion, the Second Temple is said to have been destroyed, fulfilling his prophecy that "not one stone would be left upon another" after his Resurrection. However, most historians accept that the Second Temple was destroyed some years later by Roman invaders.

Based on religious importance, the Jews have by far the strongest claim to Temple Mount. Based on historical custodianship, the Muslims' claim is stronger, having preserved and defended the area for most of the past 1,400 years. Muslims consider Temple Mount to be their third holiest site, after Mecca and Medina.

The fact that Temple Mount is now wholly within the borders of the state of Israel yet continues to be administered by an Islamic authority is profoundly, inspiringly, to the credit of the Israeli authorities and those who support them. True, any attempt to formally appropriate that land would likely lead to chaos and bloodshed on a massive scale. Indeed, Ariel Sharon's visit to Temple Mount in September 2000, during which he made some incendiary remarks about its future, is widely believed to have sparked the Second Intifada, or Palestinian uprising, and also to have facilitated, convolutedly, Sharon's own election as prime minister.

Nonetheless, the enlightened self-interest shown for the most part by the Israeli powers that be, political and religious, redounds to the benefit of all humankind. Hats off to the diplomats who finessed the situation after Israel took East Jerusalem in the June 1967 Six-Day War. And crash helmets on if Pat Robertson and his burgeoning group of fundamentalist allies succeed in disrupting this delicate balance.

MESSIAH(S) ARE COMING

Christian, Islamic, and Jewish doctrine all agree that the Messiah will physically visit the Earth one day—first stop, Temple Mount. The three faiths of course differ as to the nature of the Messiah and what he will do there. Jewish doctrine holds that this will be the first coming of Mashiach (messiah), a mortal with an additional, divine soul, who will sit on the newly rebuilt throne of David. Christianity holds that the Messiah, Jesus Christ, the Son of God formerly incarnated by the Virgin Mary, will return to occupy David's throne.

Muslims await the return of the Mahdi. In Islam, there are widely varying descriptions of who the Mahdi is and whence he comes, in large part because there is no mention of a messiah in the Quran. Sunnis, who constitute the majority of Muslims, generally hold that the Mahdi will be a descendant of Muhammad, and of Muhammad's daughter, Fatima. It is important to remember, however, that Muhammad, while considered the last and greatest of the prophets, is held to be human, not divine.

Shiite Muslims, who, like Evangelical Christians and Orthodox Jews, are an activist minority of their faith, believe that the Mahdi, otherwise known as the Twelfth Imam, Muhammad ibn Hasan, disappeared in the ninth century, at the age of five. President Ahmadinejad is among those Shiite leaders said to believe that the Mahdi's return is imminent, and that all good Muslims should do what they can to hasten that return, even if that means precipitating warfare harmful to Iranian citizens. Because ultimately, to this fanatical mind-set, the glory of the coming of the Mahdi is worth whatever price in blood has to be paid.

The Mahdi will make his appearance after a period of chaos, war, and pestilence, much as described in Revelation. Depending on the version of the story, he will also claim Temple Mount. And following that basic script, the Mahdi will lead good to victory over evil, embodied by Dajjal, essentially the Muslim version of the Antichrist, in an all-consuming world war. In fact, many Muslims expect the Mahdi to collaborate with Jesus Christ, after Jesus defeats the Antichrist/Dajjal.

Those desiring to hasten the (first or second) coming of the Messiah/ Mahdi/Mashiach agree that certain conditions must be met, the most important of which is the construction of the Third Temple on the spot where the first two temples were built. However, Islamic theologians disagree with their Christian and Jewish counterparts, who hold that the rebuilding of the Temple would require destruction of the Al Aqsa mosque, which abuts the western wall, and is all that remains from the Second Temple. A number of attacks on the Al Aqsa mosque have been launched in recent years. In 1969 Michael Dennis Rohan, an Australian, tried to burn it down. What made this attack all the more notable was that Rohan was a zealous follower of Herbert W. Armstrong, founder and leader, with his son, Garner Ted Armstrong, of the Worldwide Church of God. The Armstrongs were among the first to make extensive use of mass media to deliver their religious message; in a famous news photograph, Rohan appeared with a copy of *Plain Truth,* the Armstrongs' magazine, rolled up in his pocket. The plain truth, as they saw it, was that the Muslims had to be evicted from Temple Mount, and their structures destroyed, so that the Third Temple could be erected. At which point, Jesus, the Messiah would return. And Armageddon would begin.

Much as the politics of Ronald Reagan moved from being a right-wing fringe philosophy to being adopted by the right-of-center mainstream, the Armstrongs' preachings has been absorbed by the Evangelical movement in the United States and elsewhere, for the very good reason that it's strict constructionist, for the most part a close, literal reading of the Bible. If Armageddon is God's will, so be it. But let's not hasten death and destruction on the chance of later redemption. The eviction of the Muslims from Temple Mount, and the subsequent construction of the Third Temple, would lead to chaos and bloodshed on so grand a scale as to make today's Middle East conflict look like a school-yard fistfight. Armageddon, or some hideous facsimile thereof, will come, regardless of whether God participates or not. But then again, what better way to debut Israel's new Christian theme park, scheduled to open in 2012?

THE MESSIAH IS HERE

On October 13, 2005 (Tishrae 10, 5766, according to the Hebrew calendar), during prayer services for Yom Kippur, the holiest day of the Jewish year, the Rabbi Yitzhak Kaduri, the most renowned kabbalistic elder in Israel, lowered his head and entered a trance that lasted forty-five minutes. Many of his followers thought Kaduri, 105, was suffering an attack. When at last he opened his eyes, the rabbi announced with a beaming smile that "With the help of God, the soul of the Mashiach has attached itself to a person in Israel." When the soul of the Mashiach attaches itself to a person, that does not necessarily mean that said person will actually become the Mashiach, only that the person is a candidate for messiah-hood. But the odds are good.

"The Mashiach is already in Israel. Whatever people are sure will not happen is liable to happen, and whatever we are certain will happen may disappoint us. But in the end, there will be peace throughout the world. The world is *mitmatek mehadinim* [becoming sweet from strict justice]," declared the revered elder.

Kaduri had been waiting to see Mashiach (a transliteration of the Hebrew term for *Messiah*) for almost a century, ever since he was a boy, when the legendary Rabbi Yosef Chaim—also known as Ben Ish Chai—of Iran declared that Kaduri would live to see Mashiach. Another luminary, Rabbi Menachem Schneerson, the beloved Lubavitcher leader from Brooklyn who died in 1994, also publicly predicted that Kaduri would live to see Mashiach. Kaduri passed away in January 2006, never having physically seen Mashiach, but perhaps having glimpsed Him in his revelation.

By all accounts, the necessary precondition for the coming of Mashiach is, in addition to the construction of the Third Temple, the return of the Jews to the Holy Land. Over the centuries, many great and learned rabbis have mistakenly proclaimed that Mashiach was about to return, and that end-time was near. However, before the state of Israel was established in 1948, any such predictions were based on the very big assumption that, in the interim, the Jews would somehow acquire a homeland they could return to. To the best of my knowledge, however, no holy man of Rabbi Kaduri's stature has made such a proclamation about the Mashiach in the time since Jews returned to their spiritual home in Israel.

Sticklers insist that the First Testament demands that, before the Mashiach can appear on Earth, every single Jew in the world must go to Israel. This voyage is known as "making *aliyah*," a Hebrew term literally meaning to "go up" (the *al* in *aliyah* is the same root as the *al* in El Al airline) and used figuratively to mean ascending to a higher level by moving to Israel. But the emerging consensus seems to be that once all Jews who wish to return do so, including those individuals who might need financial or other assistance, then the condition for Mashiach's return will have been fulfilled.

For the most part, Israel's leading rabbis have been careful not to call publicly for Jews of the world to return to Israel, and instead have framed the decision to make aliyah as one for each person to decide privately. They fully realize that Armageddon is in many ways capable of becoming a self-fulfilling prophecy: If the Jews of the world were to start moving en masse to Israel, a war, whether or not of God's will, would probably break out. Beyond the destabilizing physical and economic impact, such a mass migration would almost certainly be seen by neighboring countries as a threat of biblical proportions. In anticipation of Mashiach's arrival, however, Kaduri issued the call of return.

"This declaration I find fitting to issue for all of the Jews of the world to hear. It is incumbent upon them to return to the Land of Israel due to the terrible natural disasters which threaten the world.

"In the future, the Holy One, Blessed be He, will bring about great disasters in the countries of the world to sweeten the judgments of the Land of Israel.

"I am ordering the publication of this declaration as a warning, so that Jews in the countries of the world will be aware of the impending danger and will come to the Land of Israel for the building of the Temple and the revelation of our righteous Messiah."

Kaduri's predictions, if correct, would mean the end of the world as we know it, with reunion with God for some and death and perdition for everyone else. His pronouncements, therefore, have stirred a great deal of comment. Indeed, there was some question as to whether Rabbi Kaduri actually made such a claim, but the statements have been verified.

Cynics observe that Rabbi Kaduri was a very political man, allied with the extreme right-wing religious Shas party, which apparently brooks no compromise with the Palestinians, or Arabs in general, and which favors hard-liners such as former prime minister Benjamin Netanyahu. In October 2004 Kaduri was one of the principal conveners of the Jewish religious tribunal known as the Sanhedrin, held for the first time in nearly 1,600 years.

The Sanhedrin group of seventy-one rabbinical scholars was composed mainly of supporters of Meir Kahane, a right-wing terrorist best known as founder in the United States of the Jewish Defense League, which had the slogan "Every Jew a .22," which I well remember because in college I dated a Jewish girl whose father was one of Kahane's patrons. In Israel, Kahane founded the Kach party, banned by the Knesset (Israeli parliament) as racist.

The Sanhedrin's first order of business is to rebuild the temple in Jerusalem. Kahane tried at least once, and in 1980 he was sentenced to six months in prison for plotting to destroy the Al Aqsa mosque. His spirit lives on among the current members of the Tribunal, a number of whom are associated with groups implicated in various attacks on Muslim control of Temple Mount. Charges by the Waqf Islamic authority that various Jewish groups have been tunneling in and around the Temple Mount, weakening the foundation of Al Aqsa, are met with countercharges from Jewish groups that Muslims have also been weakening the Wailing Wall and have in fact been destroying ancient Hebrew artifacts.

In the Middle East, of course, there's always a base level of this sort of dreadful activity, on all sides of the political equation. It's like an active volcano that periodically spills out some lava, perhaps dissipating pressure, perhaps readying itself for one big blow. The point is not so much to quiet the rumbling Temple Mount volcano as to prevent anyone from dropping a bomb down its maw.

Rabbi Kaduri spoke as though narrating the unfolding of history divinely ordained: "According to the writings of the Vilna Gaon, a sign of the Gog and Magog war is its breaking out on the Jewish holiday of Hoshana Rabba [the seventh day of the Sukkot holiday], just after the conclusion of the 7th shmitta [agricultural sabbatical] year."

Rabbi Eliyahu, the Genius of Vilna, whose name was discovered in the Book of Genesis by the Bible code researchers, is a legendary rabbi whose predictions, à la Nostradamus, are well revered though complicated to decode. Magog, and its king, Gog, are held to be Israel's final enemies. As described in Ezekiel 38 and 39, the end of the war against Gog and Magog also sets the stage for the ultimate Armageddon conflict to begin. (For much of the Cold War era, the Soviet Union was believed to be Magog, but that doesn't seem to have panned out.) As it happens, the United States began bombing Taliban and Al Qaeda forces in Afghanistan just after sundown on October 13, 2001, precisely the Hoshana Rabba war Rabbi Eliyahu of Vilna was talking about, and during a shmitta year, to boot. The Gog and Magog conflict is prophesied to last for seven years, at which time, autumn 2008, a major revelation concerning Mashiach will be made.

Kaduri also reminded us that, according to the Midrash, the collected commentaries of Talmudic wisdom, one of the signs that Mashiach is coming is a warming of the Earth.

When a 105-year-old sage pours his heart and soul into prophecy, one ought to consider it respectfully. If Kaduri had been content to behold Armageddon unfold, offering spiritual support and guidance along the way, we all may be the richer and indeed the healthier for it. But if his followers take it upon themselves to precipitate events, acting as though they know the mind of God, then that act of aggression against peace and stability must be neutralized.

I might have been tempted to cross my fingers and consign all the predictions of the honorable Rabbi Kaduri, and his even more estimable antecedent, Rabbi Eliyahu of Vilna, to the Talmudic arcana bin, were it not for the concurrence of another kabbalist, at the extreme opposite end of the cultural scale.

Joseph Michael Levry, a Kabbalah scholar of what can only be called the New Age variety, sees the same basic set of events unfolding, though he describes this scenario in much different ways. Based in New York City, where he founded the Universal Healing Center, Levry travels almost constantly throughout the United States, Europe, and Israel, teaching a synthesis of Kabbalah and Kundalini yoga. He says that in 2004, the world entered a period known as the Flood, the Descent of the Clouds. This period of turbulent transition, thus far coinciding with the war in Iraq and Katrina and the other megastorms, will, according to Levry, continue to be a time of intense conflict that culminates in 2012.

"The Earth, also, is fighting for survival. Indeed the world will [have gone] through eight years of purification, a sort of planetary near-death experience, through the bitter experience of natural catastrophe and/or warfare," declares the kabbalist. "The old world will be laid low in order to make way for the building of the new world of spiritual, collective consciousness with universal love at its core. The political map will be altered. There may even be a change in the geophysical stability of the world. All will come to understand that the new age that emerges, along with the devastation that came before it, was a necessary purging so that humanity could transform."

Levry sees 2012 as the year when a new and elevated consciousness will dawn: "Humanists stand at the dawn of 2012, and feel inside their heart the grip of change, the twists and turns of destiny that lie within their grasp that beckons and tempts them to follow. And now, more than ever, humankind needs all the help it can get to evolve into men and women of light."

RESIST THE PROPHECY

Armageddon just feels like it's meant to be. But those feelings have been inculcated by over 3,000 years of scripture and preaching, meaning that few of us are immune to this indoctrination, regardless of what we believe our beliefs to be. We have been given a scenario and, height of irony, we somehow feel more secure in its unfolding, even toward its megadeath climax, than in challenging or disproving the basic assumptions of Judeo-Christian-Islamic civilization. The promise of everlasting salvation is of course a heady enticement. But on balance, most of us who march to Armageddon will do so, I believe, because we've been programmed to think that way, because we've been taught that's the way life is supposed to play out.

Would God—not just the biblical character but the true, loving God, should He/She indeed exist, as I for one most gratefully believe—really desire that all who remain non-Christians somehow be consumed as hapless extras in the Armageddon Bible story? Evangelicals retort that all human beings will be given the opportunity to know and accept Christ. But the billions who remain faithful to their own sacred traditions are not going to just throw themselves into the fire pit, no matter how out-of-control the Middle East gets.

The people of the Middle East and their adherents have been extraordinarily successful at casting their own fate as the fate of the world, but the fact remains that the majority of the world's population, including but not limited to those living in China, India, and other non-Christian, -Jewish, or -Muslim countries, quite naturally feel differently. The Western world has got to kick the Middle East habit. True, much of the world's oil comes from the Middle East, but they need to sell it as badly as we need to buy it, so that all can be managed. Trillions in profit potential ensure that, one way or the other, the oil will flow.

The addiction we need most urgently to break is our slavish codependence on the region's horror and perversion, injected daily into our homes. We must understand that some dark collective need to prove that our particular deity has the biggest—do I have to say it?—is being falsely fulfilled by the Middle East melodrama, and so rid ourselves of the superstition that that conflict has a stranglehold on the fate of humanity. But sometimes it's easier to live with primitive, frightening beliefs than to calmly and maturely accept responsibility.

Easier said than done. President George W. Bush took office with the laudable intention of deemphasizing the Middle East in United States government policy. Instead, the region, particularly Iraq, absconded with his administration. A comparative handful of Middle Easterners go berserk, and summits are hastily arranged. Meanwhile China sucks the natural resources out of the Southern Hemisphere and spews the worst pollution in history with barely an eyebrow raised.

Could there actually be a preordained plan, as foretold in Isaiah and Revelation, and echoed in the Quran, that the whole thing really will come to a head in Armageddon, and that if you don't side with Jesus, Mahdi, or Mashiach, or if you do but you also believe in certain forms of world government, you are an enemy of God and therefore will explode? Sharon's incapacitation and Ahmadinejad's election are the latest in an endless series of events that, like Yitzhak Rabin's assassination, the discovery of the Bible codes, and the unearthing of the Armageddon Church in Megiddo, fit all too plausibly with the inexorable unfolding of the doomsday scenario.

Those who believe we deserve the Apocalypse because we are evil and must be destroyed, because we need redemptive violence in order to reunite with the Almighty, because the Antichrist/Dajjal is upon us, or for any other benighted reason are frightening not just because of their bloodthirsty ideology but because of the self-righteousness that drives their prophecy to become self-fulfilling. They can't wait for the ultimate fireworks display, and given a match and some access, they won't.

13

2012, THE STRANGE ATTRACTOR

"We'll all be even-Steven—that's how you say it, right?—if you do one thing," Carlos Barrios winked. Carlos, Gerardo, and I have just had a testy conversation about what the Maya see as five centuries of pitiless domination, persecution, and extermination of indigenous peoples by conquerors from the north. For reasons that can best be described as knee-jerk patriotism, I found myself defending policies that I neither supported nor even knew much about. Just then Carlos got a call regarding his and Gerardo's upcoming trip to Tokyo, where they were to conclude negotiations for providing daily Mayan horoscopes to Japanese cell phone users. Carlos shouted instructions to the lawyer, hung up, and refocused: "Stop José Argüelles!"

Stop José Argüelles? Argüelles was initiator of the Harmonic Convergence celebrations of 1987. He is the person most closely associated around the globe with Mayan science and culture. More than anyone else, Argüelles has publicized the importance of 2012. Why on earth would Carlos want to stop him?

Problem is Argüelles is crackers. Crumble him up and put him in your soup.

Consider his prediction for 12/21/12. After the galactic synchronization crews deployed at the planetary light-body grid nodes have received their orders from the Galactic Federation, advance units of the Council of Solar-Planetary Affairs will swing into action.

"The unique moment, the moment of total planetary synchronization, 13.0.0.0.0. [12/21/12 in our calendar] . . . will arrive—the closing out not only of the Great Cycle, but of the evolutionary interim called Homo Sapiens. Amidst festive preparation and awesome galactic solar-signs psychically received, the human race, in harmony with the animal and other kingdoms and taking its rightful place in the great electromagnetic sea, will unify as a single circuit. Solar and galactic sound transmissions will inundate the planetary field. At least, Earth will be ready for the emergence into interplanetary civilization," writes Argüelles.

At that moment an iridescent rainbow of collective human consciousness will arc from pole to pole, and in a single multicolored flash we will all be projected into the blissful beyond.

Wow!

Maybe Argüelles is a modern mythmaker, like Gene Roddenberry and George Lucas. According to *Hamlet's Mill: An Essay Investigating the Origins of Human Knowledge and Its Transmission Through Myth,* most great myths start out as stories about the sky. In this classic scholarly tome, Giorgio de Santillana, a professor of the history and philosophy of science at MIT, and Hertha von Dechend, a professor of the history of science at the University of Frankfurt, sift through damn near every myth ever propagated, from Amaterasu (the Japanese sun goddess who banished her brother from the sky after he threw the hindquarters of his stallion at her) to Zurvan Akarana (the mighty Iranian god of time who stands upon the world egg with an itchy copper hammer). Although at times this encyclopedic exegesis seems like a trip down Maelstrom (the grinding Nordic river that leads to the land of the dead), their thesis that mythology springs from astronomy is actually very sensible.

Imagine living in a primitive society. On dark, moonless, cloudless nights, you and your friends might spend some time gazing up at the magnificent Milky Way, and over the years you just might become pretty fair amateur astronomers. The night sky would be an obvious setting for stories; the stars and particularly the planets, which shone brightly and moved about, would likely be anthropomorphized and/or identified with great snakes, lions, or horses. What better projective test? The stories that survived centuries of yarn-swapping would become the essential myths of your culture.

When Argüelles first beheld Mayan cosmology, or what he perceived as such, he responded creatively, emotionally, associatively. Every way but literally. Which would be fine if his books were marked "fiction."

"Argüelles can tell whatever star stories he wants. But he has no right to claim that this is what the Mayans believe. You want to accomplish something with your book? Stop Argüelles! He has followers all over the world. Half a million in Australia! The book that made him famous [*The Mayan Factor*]—he wrote it without ever traveling to the Mayan world, without ever talking to the Mayan people. Mexico City [of Aztec origins] does not count," declared Carlos. He sees Argüelles as yet another usurper, more dangerous than any of the others because Argüelles is of Hispanic origin and presents himself as a native champion of Mayan culture.

"We finally met with him several years ago, and he promised to stop saying that he was talking about the Mayans, that these were only his personal theories. But then no one paid any attention to him, so he's back to claiming that his work is Mayan. Either way, the damage is done," added Gerardo.

Argüelles has lately taken to calling his work "galactic Mayan" and writes as Votan, named after a Mayan deity whose tomb was unearthed in the late 1950s. Argüelles channels Votan and then transmits these revelations to Stephanie South, aka the Red Queen. This inspiration has already taken material form as *Cosmic History Chronicles,* a seven-volume "reformulation of the human mind."

"Cosmic History is a system of thought and technique to be learned and applied in order that the human being can take the next step on the road of evolution into a holographic perceptual system," Votan transmitted to the Red Queen.

I delicately inquired of the Barrios brothers if, as Argüelles/Votan contends, Maya believe that there are antennas in the human solar plexus receiving signals from the center of the Milky Way. Gerardo's face seemed to turn into granite. Carlos popped an antacid. I persisted. Is it true, as cosmologist Brian Swimme writes in his forward to *The Mayan Factor,* that there is a beam from the core of our galaxy to which "each person has the power to connect directly—sensuously, sensually, electromagnetically," thereby bodily absorbing its energy/information? This galactic beam thing is a sore point among the Maya, because it has been used as a rationale for warnings and prophecies about 2012. Swimme summarizes Argüelles's case: "Human history is shaped in large part by a galactic beam through which the Sun and the Earth have been passing for the last 5,000 years, and . . . a great moment of transformation awaits us as we arrive at the beam's end in 2012."

No tummy antennas, no beam, the Barrios brothers confirmed. But Vernadsky, the legendary Russian planetary ecologist, might not be so sure: "Radiations from all stars enter the biosphere, but we catch and perceive only an insignificant part of the total; this comes almost exclusively from the sun. The existence of radiation originating in the most distant regions of the cosmos cannot be doubted. Stars and nebulae are constantly emitting specific radiations, and everything suggests that the penetrating radiation discovered in the upper regions of the atmosphere . . . determines the character and mechanism of the biosphere," writes Vernadsky.

Still no tummy antennas, but Vernadsky, like Argüelles, clearly believes that the biosphere, of which homo sapiens is an integral part, depends for its continued well-being upon beams of galactic radiation. So does Dmitriev, for whom "impulses from the center of the galaxy" are one of the three factors most seriously underestimated by contemporary scientists. And there is at least a poetic concurrence between this sense of galactic connectedness and the ancient Mayan belief that the Milky Way is a road of souls to the underworld, or an umbilical connection between heaven and Earth, one that will be disrupted by the Solar System's eclipse of the center of the galaxy on 12/21/12.

My former wife, Sherry, has the habit of protecting her solar plexus whenever anyone malignant or unstable walks into the room. While I doubt that her reception extends all the way to the center of the galaxy, if anyone's solar plexus could pick up signals that far away, it's hers. And maybe Argüelles's.

On balance, Argüelles's vision for 2012 is hysterical and self-serving, but I cannot shake the feeling that his bell has been rung for a reason. Did Argüelles somehow glimpse the enormity of 2012 and lose his bearings because of what he saw?

2012 AND THE I CHING

The specter of 2012 has certainly captivated some unusual minds, starting with Terence McKenna, New Age philosopher, the man anointed by the *New York Times* as the successor to Timothy Leary, who in fact once introduced McKenna as the "the real Timothy Leary." After taking his degree in ecology and conservation at Tussman Experimental, a short-lived adjunct of the University of California at Berkeley, McKenna spent a few years catching butterflies and smuggling hashish in Asia, and then with his brother David headed for the Colombian Amazon rain forest, where they "researched" something called *oo-koo-he*, the local "violet psychofluid." After a year (or was it a decade?) of tripping out on rain forest hallucinogens and grooving on the I Ching, the ancient Chinese book of prophecy and wisdom, McKenna discovered a complex fractal encoded within the oracle. He called this fractal a "timewave," essentially a repeating diagram of time's trajectory. It verifies the basic Mayan prediction that time, as we know it, will end, stop dead, on 12/21/12.

The I Ching, also known as the Book of Changes, dates back almost 3,000 years. It combines images and ideas from ancient oracles with Chinese mythology, history, and folklore. According to psychologist Carl Jung, cause-and-effect theoretical explanations are not important in the I Ching. The book's main focus is instead on the element of chance; it offers the reader a variety of ways to understand, and even exploit, coincidence by achieving a properly wholesome mental and spiritual state and by grasping the full range of the details that make up any given moment.

Did McKenna get higher than a frat house and groove on I Ching wisdom until inspiration struck? Or did he actually decipher our fate? The I Ching begins with what is known as the King Wen sequence of sixty-four hexagrams, which are drawings made up of six solid and/or broken lines. Each hexagram has its own meaning and implications, as explained in the text of the I Ching. McKenna noted that this sequence corresponds to the 384-day lunar calendar used by the ancient Chinese: 64 (the number of hexagrams) times 6 (the number of lines per hexagram) equals 384. Thus he began to form the opinion that the King Wen sequence somehow represented time. Further investigation revealed more correspondences. The average number of days in a lunar month—the Chinese have long used a lunar calendar—is 29.53. Multiplying that number by 13, the number of months in a lunar year, equals 383.89, which rounds nicely to 384, the King Wen sequence's magic number.

Convinced that the I Ching therefore represents the flow of time, McKenna set out to diagram history. Eras with high levels of innovation were represented as peaks; low levels were troughs. He discovered that the same basic peak/trough pattern repeats itself over and over again, but in shorter and shorter intervals. For example, the same graph that represents the 30,000-plus-year period from the rise of the Neanderthals to the commencement of art and music also aptly illustrates the 500-year period from when the Black Death ravaged Europe to the beginning of the Industrial Revolution. And that same set of graphics goes on to represent subsequent periods of half a century, a year, all the way down to months, weeks, days, and hours, as the timewave nears its 2012 end. Time started out as the gentlest breeze but has blown faster and faster over the course of history and is now a gale-force wind.

"For beauty is nothing but the beginning of terror," writes the German Romantic poet Rainer Maria Rilke in the *Duino Elegies*. A psychedelic mindset like McKenna's would have tripped out on that one, especially at the moment he discovered that the date he came up with as the end point of history, the day when time would finally huff and puff and blow reality away, was 12/22/12, just a day off from the date divined by the ancient Mayan prophecies.

McKenna firmly maintains that he came up with the 12/22/12 end point independently. In point of fact, the McKennas did publish their timewave theory in *The Invisible Landscape: Mind, Hallucinogens, and the I Ching* a good dozen years before José Argüelles's book, *The Mayan Factor*, thrust the 12/21/12 date into the cultural debate.

EASTERN MYSTICAL 2012

Praised by Tom Robbins, that sparkling novelist, as "the greatest visionary philosopher of our age," Terence McKenna became the darling of the Santa Fe Institute in-crowd, where concepts such as chaos and catastrophe are on speed-dial. Before he died in 2000 at age fifty-three, of a brain tumor his doctor said had nothing to do with the copious amounts of psychedelics he had been ingesting since he was a teenager, McKenna even cowrote a book, *Trialogues at the Edge of the West: Chaos, Creativity, and the Resacralization of the World,* with two Institute heavyweights. Chaos theorist Ralph Abraham specializes in determining the conditions under which organized systems devolve into anarchy, such as, say, the collapse of the global ecosystem due to an internal or external stress. Natural philosopher Rupert Sheldrake argues that Nature has something akin to a universal memory bank that her creatures tap into from time to time, hastening learning and evolution—a theory that fits neatly with McKenna's timewave of accelerating change.

In 1987, the *New York Times* asked me to explain why Sheldrake left his cushy position as a tenured biochemistry professor at the University of Cambridge and moved to an ashram in southern India where he wrote a book, *A New Science of Life: The Hypothesis of Formative Causation.* In a nutshell, Sheldrake contends that if a group of rats, for example, were taught a set of tricks in Los Angeles in April, a group of the same species of rat in, say, London would learn the same set of tricks quicker than the Los Angeles rats had. This would be the case, Sheldrake maintained, even if there had been no communication between the rats, between the people teaching the rats, or by any other known form of information exchange between the groups.

Nature having a mind strongly implies that She is some sort of sentient being, a theory that might be fine for those flakes over at the divinity school, but as far as scientists were concerned, Sheldrake should have been burned at the stake. The only one of Sheldrake's colleagues at Cambridge who would go on the record favorably about his work was Brian Josephson, who had won the 1973 Nobel Prize for a page-and-a-half monograph on what became known as the Josephson junction, an aspect of quantum mechanics that thirty years later led to the development of superconductors. At the time, Josephson was deep into the study of how the universe folds in upon itself; he ventured the opinion that maybe Sheldrake had somehow wriggled into one of the folds.

Nice, but not enough to build a 4,000-word profile on, so I finessed an extra $500 out of my expense account and got myself to India. There I discovered, quite to my surprise, that I was a dead ringer for Rajiv Gandhi, India's prime minister at the time. Extraordinary bursts of hospitality resulted from this resemblance. In one village along the Cauvery River, I was showered with jasmine wrist-wreaths and presented with a flag, in the form of a butterfly tied to a string.

Shantivanam, the Hindu-Christian ashram outside of Madras, now Chennai, where Sheldrake wrote his book, was run by an unforgettable man, Father Bede Griffiths, an Oxford-educated Benedictine sage who, before his death in 1993, wrote a number of books merging Hindu and Christian spirituality. It was easy to see how Griffiths had influenced Sheldrake. He communicated to the young scientist the Hindu sense of how important it is to understand the unseen world. Of course there is a universal mind, Griffiths preached, that's where the rats, the people, the plants, everything comes from. The universal mind is what's real, and the physical world is its greatest manifestation.

"From the beginning of history, as far as one can tell, [humanity] has recognized behind all the phenomena of nature and consciousness a hidden power . . . There is not a particle of matter in the universe, not a grain of sand, a leaf, a flower, not a single animal or human being which has not its eternal being in that One, and which is not known in the unitive vision of the One. What we see is the reflection of all the beauty of creation through the mirror of our senses and our imagination, extended in space and time. But there in the vision of One all the multiplicity of creation is contained, not in the imperfection of its becoming but in the unity of its being," writes Griffiths.

I have always found the subject of oneness particularly confusing. Is oneness type A—sixteen seconds to go, Green Bay Packers versus Dallas Cowboys, NFL championship game, Packers down by 3, have the ball on the Cowboys' three-yard line, no time-outs left, ten degrees below zero, Packer QB Bart Starr barks, Cowboy lineman Jethro Pugh snorts, fans hold breath. Is that the oneness they're talking about, the heart-pounding moment of delirium, anticipation, and focus pounding our hearts forever?

Or is oneness type B, where we all transcend such divisive trivialities as who wins or loses a football game, and know better than to care, when the whole world was expecting a pass, either to win the game or at least, if it were incomplete, to stop the clock long enough to kick a field goal and tie up the game, that Starr, not much of a runner, keeps the ball and actually steps right on the back of his guard, Jerry Kramer, to wedge himself over the goal line and score the winning touchdown, right in Pugh's ugly face?

Maybe there's a type C, a Hegelian synthesis of caring and not caring, like each of us rooting our heads off for the best possible game? The Hindu word for Griffiths's study of the true reality behind the physical one turns out to be *maya*—a word that is now also fast becoming Hindu for "impudent indigenous upstart." The Indians are the Super Bowl champs of cosmic consciousness and aren't about to be upset by no Yucatán yahoos. Indian culture has lots of money and prestige tied up in its philosophy, and to have somehow missed out on the fact that Time will end in 2012 would be downright embarrassing.

Hindu scholars date the beginning of the current age, known as Kali Yuga, to the day of the death of Lord Krishna's physical body, at midnight on February 18, 3102 BCE, startlingly close to the Mayan starting point of August 13, 3114 BCE. At the end of Kali Yuga, or the Degenerate Age, Kalki, the Hindu equivalent of the messiah, will come. Kalki is the tenth and final avatar (incarnation) of Vishnu, one of the three aspects of the Hindu supreme deity; Vishnu, Brahma, and Shiva make up what is sometimes called the Hindu trinity. Kalki will bring justice to the iniquitous and usher in a new golden age. However, that golden age is not supposed to start until the year 428,898 CE, so no one has been too concerned.

That is, not until Sri Kalki Bhagavan arrived on the scene several years ago and founded his own ashram, also outside of Madras/Chennai, not far in fact from the spot where, in 1991, Rajiv Gandhi was presented with a gift. After the explosion, all they could find were his tennis shoes.

Kalki, as he likes to be known, proclaimed himself the tenth and final avatar of Vishnu and announced that the golden age will start, after much pain and turmoil, in 2012. Kalki gives credence to the Mayan prophecies, much to the displeasure of the Brahman spiritual establishment, which has supported a lawsuit for fraud against the former insurance company clerk. The lawsuit has made it all the way to the Indian Supreme Court, but Kalki is undaunted. With the help of more than a million followers worldwide, many of whom stream through his rapidly expanding compound, he and his wife, Amma, run the Golden Age Foundation, the Oneness University, and are building the Oneness Temple, said to be the largest unpillared structure in Asia. His broadly ecumenical Global Oneness Web site is one of the largest on the World Wide Web.

Kalki ties his 2012 prediction to the transit of Venus. Venus transits the Sun, that is to say, crosses its face as seen from the vantage point of Earth, less than twice a century. The last time it did so was for a six-hour period on June 8, 2004, and it will do so again on June 6, 2012. The most recent previous Venus transits occurred in 1874 and 1882. Virtually every cosmological system accords some special status to Venus. In Mayan astrology, the 260-day Cholqij calendar is timed to approximate a woman's pregnancy and also the number of days Venus rises in the morning each year. And Mayan astronomers as early as 400 BCE had determined that Venus's synodic year is 584 Earth days, commendably close to the 583.89 Earth days we now judge it to be. The synodic year is the time it takes for an object to reappear at the same point in the sky, relative to the Sun, as seen from the Earth. Oddly, Venus's day takes up to 243 Earth days, almost half of its synodic year, to complete. The ancient Mayans believed that Venus embodied their supreme deity of goodness, the feathered serpent known as Kukulcán.

In Vedic mythology, shared by Hindus and Buddhists, Venus is called Shukra, Sanskrit for "semen." Considered to be an effeminate man who has learned how to fight the gods, Shukra gives his name to Friday, the sixth day of the week. Thus, in Hindu numerology, he governs the number six. As it happens, the next transit of Venus/Shukra will occur on 6/6/12 (6+6).

Just enough of a coincidence to cause a few double takes. But 2012 is that kind of year.

AROUND THE WORLD IN 2012

The 2012 apocalypse date originated with the disappearance of Atlantis, according to Patrick Geryl and Gino Ratinckx, who energetically prosecute this startling contention in *The Orion Prophecy*:

> The day Atlantis sank under the waters—27 July 9792 BC—Orion, Venus and a few other stars and planets occupied some "code positions." Those high priests that escaped the cataclysm took their knowledge with them and stored it in the labyrinth (The Circle of Gold) in Egypt. And right there the master plan was drafted to warn mankind of the next cataclysm. This incredibly shocking story needs to be known all over the world. Because in 2012 the stars are in the exact same position as the year Atlantis went down.

Who knows if Plato and the others who wrote of Atlantis were correct in believing that it ever existed? And even if it did once upon a time sink slowly in the east, there's no guarantee that those bubbles were actually pearls of wisdom. Geryl and Ratinckx devote much of their book to reconciling Mayan and Atlantean prophecies, which reach us via the ancient Egyptians, they maintain. But how did those ancients communicate? Through Sheldrake's universal memory bank?

The transatlantic seafaring required to travel from Egypt to Central America could possibly have been accomplished by the Phoenician sailors, though they would have been transmitting what by then had become ancient knowledge, since the Phoenicians came thousands of years after the Egyptian heyday. Or it's possible that the connection could have been made much earlier. Ancient travelers could have proceeded from Egypt, northeast across Asia, and on through Siberia to the Bering land bridge, a land mass up to 1,000 miles wide that, geologists concur, traversed what is now the Bering Strait as recently as 10,000 years ago. Paleogeneticists believe that Native Americans trace their distant ancestry to prehistoric Asians who crossed the Bering land bridge.

The ancient Egyptian voyagers would then have gone southeast across the North American continent to Central America, where they would have encountered the shadowy civilization of the Olmecs, who preceded the Mayans by many centuries, perhaps millennia.

An archaeological site known as Cuello, located in what is now northern Belize, is one of the world's earliest known settled communities. It was continuously inhabited from 2500 BCE until the end of the Classic Maya period, approximately 1000 CE, according to Thor Janson, a local scholar/explorer who has lived in Guatemala since the late 1980s. Janson has made something of a career of pointing out the similarities between Classic Maya civilization and its counterparts in Egypt, India, and elsewhere.

"Maya insignia almost precisely duplicated those of the Old World: fan bearers, scepters, tiger throne, lotus staff and lotus throne, canopies, palanquins, and the blown conch shell as royal trumpet. Astonishing similarities also exist in the basic content of the mythologies of these supposedly isolated-from-one-another Old and New World cultures. Common mythic figures include the cosmic tree of life having a bird with outstretched wings at its summit and a serpent at its roots, the four sacred colors, the four sacred directions, the four primal elements (fire, water, air, earth)," writes Janson.

Gerardo Barrios was skeptical of any Egyptian-Mayan connection until he spent several weeks exploring the Egyptian pyramids with a local anthropologist. He came back with a zip-file of photos of very Mayan-looking hieroglyphs, including several of feathered-serpent creatures that closely resemble Kukulcán. No black jaguars, though.

Not surprisingly, the 2012 end-date is common throughout Native American cultures. In *The Cherokee Sacred Calendar: A Handbook of the Ancient Native American Tradition,* author Raven Hail, a member of the Cherokee Nation of Oklahoma, enables her readers to calculate their natal days and thus learn their position in Native American astrology. The ephemeris starts on January 11, 1900, a day known as Day 1 Rabbit, and proceeds to identify the Day 1 of every thirteen-day "week" (similar to the Mayan Cholqij calendar) for the next 112 years, ending, without comment, on Day 4 Flower, December 21, 2012. Flower, Hail informs us, is the most sacred of the twenty day signs, because it is the end of the cycle. Of the thirteen energies, she writes, "Four is the most sacred number: as the four quarters of the Earth, the Four Seasons, the four phases of human life (Maiden, Mother, Mage, and Midnight)."

The Cherokee perspective fits with that of the Q'ero Indians of Peru. In *Keepers of the Ancient Knowledge: The Mystical World of the Q'ero Indians,* Joan Parisi Wilcox, an initiate into Q'ero rites, reports simply that tribal lore describes the period from 1990 to 2012 as the Age of Meeting Ourselves Again, at the end of which, Time will cease.

December 21, 2012, is also the magical date for the Hopi of Arizona. "The Hopi prophecy is an oral tradition of stories that Hopis say predicted the coming of the white man, the world wars and nuclear weapons. And it predicts that time will end when humanity emerges into the 'fifth world,' " writes Richard Boylan in *Earth Mother Crying: Journal of Prophecies of Native Peoples Worldwide.* The Hopi jealously guard their prophecies from the general public, to the point of sometimes taking legal action against those who reveal them. However, it is known that the Hopi calendar is basically in sync with the Mayan; both peg the beginning of the Fifth World, or Age, at 12/21/12.

STRANGE ATTRACTOR

"As we approach A.D. 2012, as if there is a strange time-attractor in the sky, we feel this pull instinctively. Like caterpillars undergoing metamorphosis and eventually becoming butterflies, we may actually be time-coded to change into new forms," writes Barbara Hand Clow, an astrologer and ceremonial teacher.

Whether it's a universal DNA time-code or just a bandwagon effect, Clow captures the fact that the year 2012 is fast becoming the doomsday deadline, with all manner of end-time predictions being accelerated to fit the Mayan timetable.

Take, for example, the role of 2012 in the ancient legends of the Maori, the indigenous people of New Zealand. Maori mythology foretells the reunification of Rangi (the Sky) and Papa (the Earth), a couple so tight that they crush their children between them. After years of struggle, the children, who represent humanity in this myth, finally manage to push Rangi and Papa apart, but then set to fighting among themselves. When the children are completely consumed by their bickering, Rangi and Papa reunite, destroying everything and everyone in between. Prior to the final destruction, a great canoe slips from the heavens and gathers up the relative handful of people who have managed to preserve their spiritual nature.

The reunion of Rangi and Papa will commence in 2012, according to Maori elders, who declined to be quoted directly. Their reluctance to go on record stems from the fact that a number of Maori myths and legends turn out to have been created in the early nineteenth century by Western anthropologists who visited New Zealand and either grossly misconstrued or outright invented ersatz tales that were gradually incorporated into Maori lore. A century and a half later, the true origins of these stories were discovered, and the Maori were faced with the dilemma of whether or not to renounce the stories that had been part of their culture for a century and a half. After much discussion and dissent, the Maori have apparently decided to retain these stories and will continue to pass them on in their tradition. But they'd sure like for it not to happen again.

So do the Maori legends actually point to an end-date of 2012? Or is the date simply something that sounds plausible to certain elders charged with keeping tradition? I can't say for sure. The lesson to be drawn here is that we are on the verge of a 2012 proliferation, in which it will indeed become the strange and powerful attractor that Clow claims. Every obscure and ancient tradition is potentially vulnerable to a 2012 "discovery." Purists may recoil, but my purpose in this book is to ferret out the truth about 2012, not control its linguistic usage. To that end, I feel a certain trepidation, a bit the way the

Xerox folks must have felt when their beloved trademark finally became so popular a synonym for "photocopy" that it lapsed into common parlance. I obviously don't own 2012, but I do have a vested interest in seeing that the term doesn't proliferate so absurdly that the very real, very scary possibilities associated with it are no longer taken seriously.

To a certain extent, this sort of thing happened with the term "Gaia." When I first started writing on the subject in 1986, there was one Gaia theory, advanced by James Lovelock and Lynn Margulis. Now there are at least a dozen speculative propositions stamped with the Gaia name, including one by a pagan named Otter G'Zell. Gaia also moves merchandise, herbs, clothing, geological surveys, tea, and so on. Fortunately, the gloom-and-doom aura of 2012 will probably keep it from becoming much of a trademark, except maybe for crash helmets.

Already I've had folks tell me in all seriousness that 2012 is when the ancient Greeks expected the world to end. That is incorrect. There is no evidence that the ancient Greeks gave a second thought to the date. But someone heard it somewhere and that made it true. People can blather on all they want, what do I care? But if that blather reaches a level where intelligent, concerned, potentially influential people simply throw up their hands and dismiss the genuine threats posed by 2012, we have all been endangered. It will be like the little boy who cried wolf, except that this time the wolf just might be at everyone's door.

OUR FAVORITE DOOMSDAY

Brian Cullman, a New York writer and composer, tells the story of one day being in a bookstore, and out of the corner of his eye spying a book entitled *How to Prepare for the PAST.* A mind-boggling notion. He seized the volume, only to find that the title was actually *How to Prepare for the PSAT,* a standardized college entrance exam. But the notion that the past is something to be prepared for inspired him to write a beguiling song, the moral of which is that one's history—personal, political, evolutionary—can drop in at any time.

Periodically the near-death look returns to fashion, and chances are that it will accompany the approach of 2012. Just as the consumptive look was perversely in style as tuberculosis epidemics raged at the turn of the twentieth century, and as black, pale, and sullen "heroin chic" surfaces now and again in certain tragically hip quarters of Seattle, New York, and Los Angeles, a 2012 doomsday chic seems in the cards. Already there are several rock bands, including Downfall 2012 out of Houston (their logo is a picture of the Earth with a lit fuse stuck in it) and Multimedia 2012 out of St. Petersburg, Russia. And a man who calls himself Dr. Paradise has been touring the world with a show called Paradise 2012, a concert of light and color patterns tuned to the specific frequencies of the chakra nodal points on the head and body.

Did Maximón, the playboy saint of Santiago Atitlán, have it right all along? Is 2012 just one big excuse to party like its 1999? The logic is pretty powerful. If 12/21/12 does indeed represent the end of time, or some awesome approximation thereof, what better way to ring out the old than with a bottle of champagne in one hand and your main squeeze in the other? Whether the moment is wonderful, horrible, or a big, fat dud, we can keep the party going through Christmas and New Year, and then sober up some time in January, when we're all fat, broke, and cold. That's when the realizations might really start slithering out of our ears. Were we partying to ward off catastrophe, or were we secretly hoping for a divine end to it all?

Margaret Mead famously remarked that she never came across a people without a creation myth. Be it the eternal mythology of Greece, the Babylonian love battle between Marduk and Tiamat, or just two lovers and a snake in a garden, people need some explanation of how everything came to be. And apparently, of how it all ends: runaway global warming, nuclear holocaust, Armageddon, and now 2012. If nothing else, the Apocalypse 2012 movement has helped focus our need for what Frank Kermode, the great and hoary Oxford literary critic, called "the sense of an ending," in his brilliant book of that name.

Those who prophesy doom usually claim to fear and loathe it and to pray that they are wrong. But I think doomsday has a profound if unspeakable allure for those who are unhappy with themselves, their society, their Maker. Accepting that doomsday is imminent provides the believer with immense satisfaction—that he or she possesses the most important knowledge in the world and that all other pursuits are trivial or misguided. It's a form of vicarious revenge that anyone can take on life's unfairness.

Doomsday also helps fill the void left by the US–Soviet nuclear holocaust paradigm that haunted our collective imagination until the late 1980s, a void curiously unfulfilled by the threat of global terrorism, which is every bit as heinous though only weakly apocalyptic. Will the threat of Apocalypse bring nations and people together, to unite against the common "enemy" of extinction? How big a scare is necessary to make people forsake their old warlike ways? The idea of 2012 serves as a conceptual bridge to the profoundly disturbing possibility that, sooner or later, life as we know it could indeed end, horribly.

At any given point in history, there has always been a chorus of eccentrics predicting the end of the world. What makes today different is the broad-scale convergence on the date 2012. How do we account for the fact that traditions as diverse as the Bible, I Ching, and Mayan, plus the swell of recent scientific evidence, indicate that it all may be coming to a head in 2012? Is there an underlying sense of doom in contemporary culture, perhaps a fearful, primitive response to accelerating globalization, finding expression in the 2012 movement? If only so much of the information about 2012 did not violate one's sense of intellectual decorum: prophecies from Mayan shamans, interstellar theories of obscure Siberian geophysicists, ruminations of South African psychics, decrees of kabbalist rabbis. Indeed, no single source, no matter how persuasive, could or should move one to seriously ponder the imponderable of the world tumultuously metamorphosing in 2012. But when such disparate cultures and disciplines come together in fundamental agreement that dramatic, deadly change is on the way, it is only prudent to pay heed and move forward together to prepare ourselves, our loved ones, and whatever fraction of the greater world we might be able to influence, for the coming events.

RATS EATING YOUR FACE. The flipside, of course, is that doomsday embodies our worst fears, such as Winston's deepest terror in *1984*. Everyone has a special fear that affects him or her disproportionately. There are two broad categories: evil and senselessness. Some find the idea, for example, of being shot and killed by a murderer far more frightening than, say, being shot and killed accidentally, the way Vice President Dick Cheney almost managed to do to his friend in that hunting accident. Pain and everything else being equal, I'll take the murderer. At least it's a relationship, not just a switch being flipped.

Most doomsday scenarios tend toward senselessness. Cynics natively grasp the perverse emotional logic of being betrayed by the Sun, Earth's warmest relationship. After all, at the human level, aren't violent crimes likelier to be perpetrated by loved ones rather than strangers?

Those more inclined toward suicide might opt for something like the self-annihilating volcanic spasm scenario and perhaps for the rebirth that invariably ensues. Fatalists shrug and say, "Whatcha gonna do?" at the prospect of the Earth being beaned by a rogue comet or asteroid. The moral of the story is that there is no moral. They just hope it's glorious, with enough awesome beauty to make the final moments mesmerizing.

2012 IS NOT PRO-DEATH

When I started my research into 2012, I expected to turn up a lot of material about recent doomsday cults, such as Aum Shinri Kyo of Japan, Branch Davidians (David Koresh, Texas), Heaven's Gate (California), Jeffrey Lundgren (Mormon), Movement for the Restoration of the Ten Commandments of God (Uganda), and the People's Temple (Jim Jones, Guyana). I found not a single such reference to 2012, in almost a year and a half of daily investigation into this topic.

The reason that doomsday cults were not found in this research is that 2012 is not about death—not from the Mayan perspective, not from the interstellar energy cloud, not from the changing Sun. It is about a major transformation that may entail a great number of deaths, human and otherwise, but unlike most of the doomsday cults, there is nothing in the spirit of 2012 that advocates death as a means to transcendence or anything else. Death is simply what will likely occur, not a recommended solution.

Still, doomsday is doomsday, and with the very real possibility that everything we know and care about may be coming to end, we need to find faith that something, anything, is coming next.

On the night I arrived in South Africa I went to dinner at the home of Pierre Cilliers, distinguished researcher at the Hermanus Magnetic Observatory. Our correspondence had indicated that Cilliers was a gracious man, and I fully expected a delightful evening of talk about pole shifts, magnetic declination, and the like. I instead found myself in the company of devout and delightful Christians, who freely shared with me their love of God and His son, Jesus Christ.

My editor once called me "the good-news, bad-news guy," always balancing, tending toward the middle. I am certainly no fundamentalist, though I do sometimes admire, even envy, the strength of their beliefs. And over the years I have noted that when strong-minded, good-hearted people believe something passionately, in a way that lights up their life, it seems that God is smiling through them. Two of Cilliers's guests, white South Africans, told how they had, with forty-five dollars and their faith in Jesus, established a bustling elementary school north of Johannesburg. They kept it running and even expanded it, despite a firebombing by white separatists who despised their integrated school. Another guest gently told how he and his wife had for years tried desperately to conceive. Finally an adorable baby was born, only to die seven months later of cancer, a disease that took his wife shortly thereafter. He explained that God is not the cause of the ills, or even the joys, of one's life. Instead, the important thing to know is that, whatever happens, God is always with us and wants only for us to turn to Him.

These folks' faith had made them stronger in the face of adversity than I could ever be. Did their faith also make them more perceptive to the truth of the Almighty?

At no point did any of the dinner guests express any desire whatsoever for Armageddon to occur, but clearly that would be all right by them. They were ready to receive God's love in whatever form it might come. I wanted to know what Cilliers thought about Revelation and Armageddon, but I waited until the next day, until he was in his office, in business/science mode, to avoid an expansive, end-of-evening response that he might wish to backtrack from in the cold light of day.

"The Lord said that a sign of the end of times would be an increase in storms, earthquakes, and other catastrophes. He is the author of the Bible and the author of Nature. When we see conflict, it is either because we don't understand his Revelation or because we confuse our observation," said Cilliers, busily pulling together his papers. He was leaving the next day for an atmospheric physics conference, where he would present his research on the relationship between solar variability and fluctuations in the Earth's magnetic field. But he paused to reference what Jesus Christ had said on the matter:

"The time is coming when you will hear the noise of battle near at hand and the news of battles far away; see that you are not alarmed. Such things are bound to happen; but the end is still to come. For nation will make war upon nation, kingdom upon kingdom; there will be famines and earthquakes in many places. With all these things the birth-pangs of a new age begin," said Jesus Christ (Matthew 24:6–8).

Genuine birth-pangs increase in frequency and intensity as one approaches the final blessed event. The question is, are the wars, terror, famine and diseases, hurricanes, earthquakes, and volcanoes of the past century, and particularly since the dawn of the new millennium, real labor contractions, or are they just Braxton Hicks warm-up pangs? Are we even pregnant? Or is this simply a case of massive indigestion?

I asked Cilliers what he thought about the possibility of the world, as we know it, coming to an end, or profoundly, abruptly changing, in 2012.

"It is not unlikely that this will occur in our lifetime," the sixtyish geophysicist replied.

LISA WILL SHOW US THE WAY

Mother-Father-Child. Father-Son-Holy Spirit. Vishnu-Brahma-Shiva. Thesis-Antithesis-Synthesis. The Lover–The Beloved–The Love Between Them. Hydrogen-Oxygen-Hydrogen. Executive-Legislative-Judicial. Body-Mind-Spirit. LISA.

Triangles, conceptual and otherwise, are sacred, none more than LISA (laser interferometer space antenna), the unfathomably enormous equilateral triangle designed by the European Space Agency and NASA to surf gravitational waves while orbiting the Sun:

> Lisa will be able to detect the gravitational shockwaves emitted less than a trillionth of a second after the Big Bang. It will consist of three satellites circling the sun, connected by laser beams, making a huge triangle in space 5m km on each side. Any gravitational wave which strikes Lisa will disturb the lasers, and this tiny distortion will be picked up by instruments, signaling the collision of two black holes or the Big Bang aftershock itself. Lisa is so sensitive—it can measure distortions a tenth the diameter of an atom—that it may be able to test many of the scenarios being proposed for the pre-big-bang universe, including string theory.

Far out! A laser beam triangle, 3.1 million miles on each side, circling the Sun, quivering with infinitesimal gravity waves left over from the first trillionth of a second of Creation. What better way to close our story, or at least this chapter of it, than to prove scientifically that the big bang, our latest creation myth, is indeed based in fact?

I hereby nominate Gregory Benford, a physicist from the University of California at Irvine, to be in charge of interpreting the data. Benford's "Applied Mathematical Theology: You Have a Message," an endpiece in *Nature* is a mind-blowing fable about scientists finding a pattern in the cosmic radiation left over from the big bang: "Spread across the microwave sky there was room in the detectable fluctuations for about 100,000 bits—roughly 10,000 words . . . But what did it mean? Certainly it would not be in English or any other human language. The only candidate tongue was mathematics."

The world's greatest physicists, mathematicians, philosophers, and theologians work together, in Benford's tale, to decipher what the message was. These great minds become more and more certain that there indeed is a message, but they can never figure out what it is. But the mere fact that there is a message—from God, the Universe, the Creator—inspires and enlightens the multitudes, energizes the economy, imbues reverence for the environment.

Benford's vision could be what the ancient Maya saw all along: the dawn of a new age of enlightenment. Maybe God will strum LISA's laser beams with His immortal song.

LISA's launch date is 2011. Science begins in 2012.

CONCLUSION

The Shehabs, my ancestors, remained Muslim, for the most part, until 1799, when Napoleon Bonaparte sent a sword, a bribe actually, to L'Emir Bashir Shehab II, a rustic emir who pretty much ruled the mountains of Lebanon. Without doubt, L'Emir Shehab was my hairiest ancestor, with a beard down to his navel and eyebrows so bushy that a sparrow could perch upon them.

Napoleon, age thirty, had decided it was time to conquer the Holy Land and was laying siege to the Ottoman port of St. Jean d'Acre, now Akko, on northern Israel's Mediterranean coast. Acre was primarily defended by the British fleet, which Napoleon figured he could handle by himself, but he needed someone to take out Al-Jazzar, the Ottoman pasha of the Eastern Mediterranean coast. Al-Jazzar was dismantling his own harbor, shallowing it out in order to muck-snare any invading boats before they reached shore. A Christian Bosnian who once sold himself into slavery, Al-Jazzar earned his name, which means "the butcher," as lord high executioner for Ali Bey, the Muslim sultan of Egypt.

So Napoleon sent L'Emir Shehab a fine jeweled sword and a note asking him to charge down from his mountaintop and stab Al-Jazzar in the back.

Victory would yield my ancestor control of the eastern Mediterranean, a fact never quite lost on this hereditarily impecunious descendant.

L'Emir Shehab accepted the gift but sat out the battle, and after a sixty-one-day siege Napoleon withdrew in defeat. France was soon driven from the region. To pay off the debts from his Middle East campaign, Napoleon, who by then had crowned himself emperor, did what he swore he would never do. He sold off the Louisiana territory to Thomas Jefferson in one big chunk, rather than dividing it among several nations to keep from creating a North American superpower.

When L'Emir Shehab got word that Al-Jazzar was going to execute him anyway, for not having helped fight off Napoleon, my ancestor took the sword, snuck down from his mountain and onto a boat for the island of Cyprus, where for the next four years he hid out in a monastery praying for Al-Jazzar to die, which he did in 1804. As the story goes, L'Emir Shehab went into the monastery a Muslim and came out a Christian, one reason Lebanon is almost 50 percent Christian today. Actually it's a lot more complicated than that, but suffice it to say here that L'Emir Shehab reclaimed his throne and, after putting out the eyes of a lot of treacherous cousins, ruled the mountains of Lebanon, practicing Christianity, Islam, and the Druse religion, a kind of Muhammadless Islam, simultaneously. Things went pretty well, and L'Emir Shehab built a grand palace, Beit Eddine, which serves as the Lebanese president's summer residence today.

Over the decades, Napoleon's sword became the object of a bitter dispute among the Shehabs. The Christian wing of the family wanted to keep it as an heirloom, but the Muslim wing despised it as a symbol of Western/Christian corruption, and they knew it would fetch a fortune. So in the early 1930s, cousin Kamil Shehab, a Maronite Christian, smuggled the sword into the United States and brought it for safekeeping to my grandmother's rent-controlled fourth-floor walk-up in Park Slope, Brooklyn. She wrapped it in an army blanket and stuck it behind the ironing board.

Mostly the sword didn't do anything more than take up space in our only built-in closet. Couldn't even hang it on the wall because burglars might hear about it. Then one day a curator from the Metropolitan Museum of Art called, wanting to hang the sword in a special exhibition. My grandmother, about five feet tall and one hundred pounds, wrapped the sword in a nice quilt, then herself in the fur coat she had borrowed from Mrs. Subt, walked down four flights of stairs and over a block to the Seventh Avenue stop of the F train, switched for the A train at Jay Street/Borough Hall, then went up the stairs and the ramp at the Broadway/Nassau station for the number 4 train to Eighty-sixth Street and Lexington Avenue, then up the stairs and over to the museum entrance at Fifth Avenue and Eighty-first, a good half mile. It wasn't that she couldn't afford to take a taxi, just that no one ever took one from Brooklyn all the way into New York.

The curator led my grandmother to the gallery where a life-sized display of L'Emir Bashir Shehab II, draped in silks and sashes but still looking like the Devil in the Tarot deck, would brandish the sword inside a glass display case. From Cyrillic engravings on the sword and some documents that were sent along with it, the curator came to be of the opinion that it was originally made for Ivan the Terrible, probably in the late 1500s.

Were this a fiction story, the sword in the display case would have glowed like thirty pieces of silver. A suave, unscrupulous museum curator would steal it and replace it with a forgery, leading to murder and Da Vinci Code revenge. No such luck. Napoleon's sword was displayed and then returned to its closet without a hitch. I honestly don't think that, in almost thirty years of safeguarding the sword, the idea of selling it ever occurred to my grandparents. After all, their rent was only forty-eight dollars a month.

It's an odd study, where people get their pride. Eight families live in the same creaky old walk-up, kill the same roaches, hang out on the same tarry roof on muggy summer evenings to catch a little breeze. Yet one family, no richer, is sublimely assured that they belong to the grand scheme of history, right up there with the asterisks and footnotes attached to the really big names. Long before knowing who Napoleon was, or why being Ivan the Terrible probably wasn't a good thing but might have been a great thing, I knew that my ancestor had been a king. As a little boy I wanted to be a king too, so every now and then when we were visiting my grandmother I'd wrestle the sword out of the closet and drag it around her living room. But sooner or later I'd get yelled at because the jewels that studded the sheath were scratching up her nice wooden floors.

So you see what I mean about being a minor character in my own autobiography. Start with the meteorite smashing the desert, color in all the big historical names, do a sword-pen-mightier-than symbolism pull-through for the highbrows, sprinkle in enough heartfelt personal stuff so the average reader can relate, and then bang home the message, say, that the whole Middle East situation is really just the nth round in the Islam v. Christianity prizefight, which is pretty much how the Muslims see things, and how we probably would too if we weren't still a little woozy from punching out the fascists and the communists. Next find a literary publisher, probably someone who wears black and smokes, get it to reviewers too PC to diss a Third World–type story, and up on the shelf for my kids' kids' kids to read goes great granddaddy, in hardcover form. Purpose fulfilled, one is allowed, though certainly not required, to die.

I've been trying to slam-dunk this thing home for the past twenty years. I'm no Michael Jordan, but it just seems that Someone up there has been playing some serious D. I guess no one is allowed to bow out of his own story, no matter how unimportant he is.

LOTS OF WRITERS are superstitious about finishing their memoir, because that pretty much translates to finishing their life. In fact, it's very hard to end any book that you care about. Subconsciously, but not too far down in the psyche, more like a whale that keeps coming up for a gulp of air and then diving back under the waves, there's the fear of letting go of what has become an intense and consuming relationship. Postbook life feels like an empty bed.

Normally this occupational hazard is offset by the prospect of receiving the rest of the advance money, paying off bills, and having a life again. But to me, finishing *Apocalypse 2012* feels like closing the book on my marriage, after which I will die, just as my father did when his wife left him. But that's all emotional bullcrap at the end of the day.

What parting words suffice the impending Apocalypse? Stick your head between your knees and kiss your ass good-bye? Ha, ha, April Fool? The next life will be a better one, God guarantees it? Grab your guns and head for the hills, the Moon, or Alpha Centauri?

NAPOLEON'S SWORD has long since been returned to Lebanon, to hang in the Beit Eddine palace that my hairy ancestor built. All that I have left is the translation of its Cyrillic engravings. Beneath the Holy Cross on the hilt are etched these declarations:

THROUGH THE CROSS THE CHURCH INITIATES ITS ACTS
AND UPON THE CROSS IT PLACES ITS HOPES.
THROUGH THE CROSS WE BEGAN OUR ENTERPRISES
AND BY THE CROSS WE SHALL ACCOMPLISH THEM.

Along the blade, two little prayers are engraved in a delicate script:

O, ye Holy Cross, be my strength and my aid
Also, be my guardian and saviour from those who wish
To attack me.
Be my armor and preserver,
Be my support and help me to be victorious.

The Cross preserves the Universe,
  The Cross is the Church's beautifier,
The Cross is the symbol of Tsars
  The Cross is the strength of the believer,
The cross is the pride of the angels and
  The chaser of devils.

On the other side of the blade, beneath an engraving of the Holy Virgin Mary, is a softer beseeching:

Upon Thee, I place all my hopes,
  O Virgin Mary, the Mother of Christ
    Keep me under the shadow of thy holy garments.

If only I could get my hands on that sword. I would slice up the billion-ton comet headed our way. Skewer the sunspots that will soon mottle the Sun's face. Stick it, like the little Dutch boy's finger, to plug up the lava flow from a supervolcano. Or maybe just grab onto it, like a security penis, and pray for the evil to pass.

PRAY

The surest way to remain safe from the 2012 holocaust is to beseech the Almighty's protection. Of course, if it turns out that there is no Almighty, or that He/She in His/Her infinite wisdom chooses not to protect us, then we're cooked. But look at it this way: nothing less than an omnipotent deity could help us out of the predicament that the 2012 prophets say we're in, so what other choice is there for us than to drop to our knees?

Over the years, I've done a lot of praying, mostly of the "Mayday! Mayday!" variety, like the time I thought I was having a heart attack but actually just got bruised in the chest from playing football. If for no other reason than good spiritual etiquette, I have learned that it is helpful to start off the prayer with thankfulness: for whatever good happens to be going on in one's life, for the fact that there is Someone up there to give thanks to, or simply for being alive and conscious enough to muster a prayer in the first place. It also strikes me as good form to ask God how He's doing, and to wish Him well.

If Christian, do not forget Mary, a nicer, more loving Mother could not exist. But don't piss Her off. As an old Italian missionary who had just survived massacres in Sierra Leone explained, Mary is the one you turn to when God and Jesus just don't want to hear it anymore. Lose Her, you roast for sure.

I've also taken to giving Gaia my regards, though we're still in that awkward stage, where I no more than half believe she exists, and if she does, she's certainly not used to human acknowledgment, much less compliments. But we're getting on. I have the distinct impression that Gaia blushes—please, dear reader, accept my apologies, if you find the following comparison offensive—in a way that reminds me of the time I impulsively gave a sweet, overweight, and bashful lesbian a hug.

Yes, I fully accept that the preceding few paragraphs may be nothing more than a description of the figments of my imagination and therefore of little or no practical worth to anyone else. But hardly anyone talks about prayer. Meditation, you can probably get a bachelor's in. Out-of-body experiences stepped into the limelight whenever it was they stopped being called daydreaming. Prayer, well, maybe it's ultimately such a personal thing that not much more need be said, perhaps other than to suggest a small prayer:

Dear God,

Thank you for the thousands of wonderful years you have given us on this Earth. Thank you for your infinite gifts of joy, love, excitement, and satisfaction, and for all the other magnificent feelings, expressible and inexpressible.

In a very short time from now, in the year 2012, a great Catastrophe may forever alter our way of life. If we, your children, must experience the fear of Apocalypse 2012 to unite in common purpose and forsake our sinful ways, so be it. We humbly accept your wisdom. But please, dear God, if it is within the scope of Your will, spare us from the actual death and destruction of Apocalypse 2012. If not for our own sake, then for the sake of those good and faithful servants who will otherwise have no chance to know your love, and to love you in return.

Amen.

OFFER UP A SACRIFICE

The pen, they say, is mightier than the sword, a claim—big surprise—made by a writer. But neither pen nor sword is mightier than the symbol: the Cross, the Crescent Moon, the Star of David, the American flag. Or the white flag, for that matter.

WE MUST APPEASE Mother Earth, cravenly and immediately.

No way by 2012 can we reverse global warming, ozone depletion, and the other ecological cataclysms already in progress. The most we can hope for is to soften their impact a bit. So we are reduced to praying that Mother Earth somehow exists in a sentient enough form to appreciate our symbolic attempts to turn over a new green leaf.

Let's start by desecrating an icon.

Humvees are certainly tempting targets; so is Arnold Schwarzenegger, who is said to own a number of those obnoxious vehicles. That the federal government for so long exempted Humvees and other sport-utility vehicles (SUVs) from the Environmental Protection Administration's mileage requirements because they were "light trucks" was an assault on common sense and commonweal depraved enough to make even the cigarette tycoons blush. It was a loophole, as few pundits could resist pointing out, big enough to drive a truck through, which is exactly what the automobile and oil companies did. Politicians who defended this sham should be defeated, prosecuted if grounds can be found, and, worse for those compulsive socializers, shunned.

SUVs, particularly those with a collapsible third row of seats, at least have the redeeming quality of being useful, able to haul lots of passengers and stuff. So those of us looking to pool our resources to lawfully acquire a vehicle and then ceremonially sledgehammer it into recyclable bits might consider the new, $101,300 Volkswagen Phaeton twelve-cylinder, four-seater luxury sedan. According to cars.com, the Phaeton ranks among the ten worst cars in the world in mileage, managing twelve miles per gallon in the city and eighteen miles on the highway, when the car is perfectly in tune, the tires inflated to specification, and premium fuel is used. Humvees are like hybrids compared to the Phaeton.

The Phaeton is the absolute, superpremium top of the Volkswagen automobile line, and smashing it to bits, in a safe and lawful manner, will send a migraine message to the world's automobile companies that it is evil to introduce gluttonous new gas-guzzler models in this day and age. It seems only reasonable when you remember where the name came from. Phaeton, in Greek mythology, was the son of the Sun god Helios. Phaeton talked his father into letting him drive the sun chariot across the heavens, but he wasn't very good at it and lost control. Seeing that he was about to crash, Zeus killed Phaeton with a thunderbolt to save the Earth from burning up.

Phaetons have the extra attraction of being made by Volkswagen, the company that produced so many excellent tanks and amphibious assault vehicles for the German Nazi regime. Volkswagen did redeem its image with the lovable old Beetle and the VW hippie van, and now the new Beetle, which comes complete with an artificial flower, several of which were clenched between the teeth of the top VW executives caught in 2005 using company funds to buy hookers, who will be offered modest honorariums to regale us at the Phaeton-smashing after-party.

Admittedly, the odds of Mother Earth grasping our symbolism are a long shot, but at this late date it may be the best shot we've got. And even if She does, Sun-based conniptions and interstellar energy clouds are beyond Her control. But symbolic acts are as much for the doer. At best they focus concentration and strengthen resolve for the serious work ahead. At the very least they impart the illusory satisfaction of having done something, when at the end of the day there may be nothing that can really be done, if Apocalypse is truly bearing down.

BLAST OFF

In trying to decipher the plot logic of 2012, to get a sense of how this story naturally plays out, the recurrent theme is "threat from above." From the threatened solar maximum, to ancient Mayans calculating eclipses of the Milky Way, to Dmitriev's interstellar energy cloud, to the angry God above descending for Armageddon, and even to the simpler threat of our satellite network being fried in the sky, it almost seems as if humanity is being beaten back, whether because we are exceeding our natural limits or because we've got to prove our mettle in order to expand into the cosmos.

My feeling is that 2012 will turn out to be some kind of comeuppance for high-tech society. We may be facing a battle with Nature, in which neither side has a monopoly on virtue. Mother Earth may well react negatively, if not necessarily consciously, to our intruding ourselves beyond our rightful sphere. Escaping from the atmosphere, Earth's outermost layer, is physically very difficult for spaceships to do; most of the rocket fuel is burned up breaking that barrier. So it stands to reason that escaping economically and culturally would also pose an extraordinary challenge. If the metamessage of 2012's threat from above is to pay more attention and take better care of the Earth beneath our feet, then point well taken. We've spent centuries violating the commonsense rule "Don't shit where you eat," and if it takes a world-threatening cataclysm for us to change our ways, well, that's what it takes. But if this is somehow Mother Earth's way of telling us not to venture beyond her apron strings, then she, like all other overly possessive parents, must be lovingly put in her place.

What better event than the 2012 catastrophe to precipitate the human colonization of space? Ideally, of course, it turns out that it was only the threat that impelled us to action; the fact that, in this optimal scenario, there was no actual devastation can be the source of enough wry humor and bemused reflection on how Life has a funny way of working out to fill the rest of the century. Regardless of which way it plays out, I like to imagine a monument to 2012, just as in Enterprise, Alabama, there's a monument to the boll weevil, which devastated the cotton crop and forced the South to diversify its economy. Better make that three statues capturing how the prospect of 2012 forced us to colonize space: one for the Johnson Space Center in Houston, one for the Moon, and one, if the apocalypse is truly Solar System–wide, for some galaxy beyond.

RAISE BILLIONS

In 1995 I wrote an op-ed piece, "Who Will Mine the Moon," decrying the lack of progress that physicists had made toward controlling nuclear fusion, despite receiving tens of billions of dollars in research funding. Nuclear fusion is probably the greatest force in the universe, the one that powers the Sun and hydrogen bombs. Controlling it is a noble goal and might one day yield almost limitless supplies of energy. It could make the burning of fossil fuels a thing of the past. Problem is, nuclear fusion is so wild, the utmost physical manifestation of $e = mc^2$, that it takes more energy to keep the reactor from exploding than the reaction produces.

My op-ed suggestion was simply that 10 percent of this funding go to promising alternative approaches to fusion control. The *Times* sent my check for $150, Dan Rather wrote a nice letter, some man who said he was an ambassador called and offered to take me to lunch but then canceled. Not a penny of funding was questioned, much less diverted. At this writing, more than eleven years and probably as many billions down the rat hole later, the plasma physicists are, if anything, farther from their fusion goals.

It's time to pull the plug.

Specifically, I propose a moratorium on all funding for controlled nuclear fusion research, effective immediately and extending through 2012. Shifting controlled fusion research funds into catastrophe prevention and relief is a good start. Each person who lives through the coming assault will likely find himself or herself in terrific need. Civil preparedness must be restored to the high political priority it enjoyed during the Cold War, when even schoolchildren were drilled in what to do in the event of an atomic bomb blast, their little psyches remaining intact. Just as Londoners took refuge in the Underground during World War II, they will do so again in case of another catastrophe, as will residents of New York, Paris, Moscow, Tokyo, and the many other major cities blessed with underground mass transit networks. Surprisingly, it turns out that there are also a number of ancient underground cities around the world that could, if properly refurbished, provide shelter. One of the largest such cities, in Derinkuyu, Turkey, originally housed up to 200,000 people.

Preparing underground networks, ancient and modern, for sudden, massive influxes is mostly a matter of ensuring structural soundness and providing adequate sanitary facilities, water, foodstuffs, medicine, clothing, and blankets. Similar provisions should be made to ensure that stadiums, high schools, auditoriums, and other public gathering places are ready for sudden, massive influxes of refugees. Imagine how much more humanely the Katrina aftermath would have been handled had the New Orleans Superdome been even rudimentarily prepared to shelter the victims.

Since the megastorms are more and more widely believed to have been pumped up by global warming, how about a little greenhouse gas tax to offset the cleanup costs, burials, and such? Global warming may not be the cause of Apocalypse 2012 per se, but it certainly aggravates our seismic and volcanic vulnerabilities, and it compounds the energy overload problems we will be facing from a tumultuous Sun and/or the interstellar energy cloud. The government/industrial complexes of China and India might pony up some of their pollution bonanza, that loophole in the Kyoto Accord that allows them to piss as much carbon dioxide into the atmosphere as they possibly can, without repercussions. The United States, the biggest gas emitter, is also cordially invited.

The money will come in buckets and dribs. I think I know where I can score 800 Gs, less certain personal and business expenses. Not long after arriving in South Africa, I was gathered up by two enterprising young gentlemen who had researched me some and had come to the conclusion that I might be of assistance in a business endeavor. They discreetly informed me that they had embezzled $4 million from their organization. For the simple favor of carrying said cash in a suitcase into the United States and then securing it in a bank safety deposit box, I would earn a commission of 20 percent.

The prospect of prison or, who knows, Guantanamo Bay, since illegal transfer of funds now seems to fall under the Homeland Security domain, made up my mind, more or less. But South African friends upbraided me later when I told them this story. They said it was immoral *not* to have taken the money, even though the funds had been embezzled, possibly from the public till. The way things work in Africa, they explained, is that all windfalls are illegal, either formally or morally, as in the case, for example, of diamonds and gold, still skewed unconscionably toward the whites. Therefore, when Fate makes the offer, one's duty is to accept, donate 10 percent to a worthy charity, and see that the windfall is otherwise responsibly deployed. Like buying us some Volkswagen Phaetons to smash up.

PREPARE OUR MINDS

How can we help defend against mass psychological collapse? Forewarned is forearmed, as the saying goes. Certainly the trauma will be lessened if we are not caught off guard entirely.

In *Embattled Selves: An Investigation into the Nature of Identity through Oral Histories of Holocaust Survivors,* Kenneth Jacobson asks the very tough question of what it took for Jews to emerge from the Nazi Holocaust psychologically whole. His first answer, much simplified, is generosity. Those who helped others cope or escape tended to have calmer, saner, postwar lives. Those who simply acted selfishly—and Lord knows, selfishness in the face of blackest evil is certainly justified—nonetheless suffered more severely later on. That generosity is not just good for you but actually essential for psychological survival is a bit of wisdom that we all would do well to absorb.

Jacobson's second key is identity, refusing to let go of who one was before the crisis began. This was particularly important in the Nazi Holocaust, because genetic identity was the deciding factor in who was persecuted, exterminated. Jews who denied their heritage in order to escape tended to suffer more later on than those who did not, though the deniers did have a better chance of living to tell the tale.

Of the hundreds of stories and remembrances recorded in this remarkable book, one epitomizes Jacobson's findings. Through slipups and subterfuge, Maurits Hirsch, a Jewish man, was mistaken for a Gentile and installed by the Nazis as mayor/administrator of a town. This man's survival instinct told him to flee before he was exposed, but he stayed because, by playing the Nazi pig, spitting on people rather than actually harming them, he could do a lot to defend and protect the people of that town. To keep from losing his sanity, Hirsch would take walks deep in the woods, and when he was sure no one was around, he would sing himself Yiddish songs.

Knowing that an apocalypse may in fact be on the way in 2012, and that generosity and embracing identity tend to minimize any posttraumatic stress disorder that results from this upheaval, we can sing our songs and face squarely even the darkest possibilities.

MAKE WAY FOR THE MAYANS

Look at it this way. If this 2012 thing pans out, and the year is fraught with catastrophe, or even scary near misses, the Mayans, having seen it all coming two millennia ago, will be in the driver's seat for centuries to come.

So between now and 12/21/12, common sense says hear them out, lure those elders out of their caves, get them on talk shows, hire them as consultants. The result, I guarantee, will be both confusing and uplifting. Confusing because of language and culture gaps, because their standards of logical rigor and consistency are inferior to ours, because our standards of metaphor and simile are inferior to theirs. Uplifting because the heart of the Mayan 2012 prophecies is about transformation, not about the catastrophes that seem destined to accompany it. They really, truly believe that 2012 is the best shot in the past 26,000 years for humanity to become enlightened and move closer to the gods.

Half-kidding, I asked the Barrios brothers what stock tips they had for investors contemplating 2012. Basics: food, shelter, clothing, computers.

"So what do we do?" I threw myself on the mercy of the court.

"The elders say we need to be returned from machines to humans," said Carlos.

"We must transform our curiosity into real purpose, of serving each other and Mother Earth," said Gerardo.

HEAD FOR THE HILLS

Jerusalem. Mecca. Angkor Wat. Tikal. Thingvellir. The Vatican. Berea, Kentucky.

Of all the sacred sites in the world, none embodies the sacred Mayan values of service to humanity and Mother Earth like the town of Berea, Kentucky. Tucked away in the Appalachians, due north in fact from the Mayan territory, Berea is charming but unassuming. The heart of the town is Berea College, which is routinely ranked as one of the top liberal arts institutions in the United States, with exceptionally high academic standards and fine moral values.

Berea College receives no federal, state, or local funding and has independently amassed an endowment of over $200 million from charitable contributions. This despite the fact that Berea students, who pay no tuition, all come from economically disadvantaged homes. About three-quarters are from Appalachia, though not even the children of alumni are admitted if their household income is above lower-middle-class level. All students work a minimum of twenty hours per week, producing pottery, wrought iron, and woodcrafts so fine that the school cannot keep up with demand. The wood is sustainably milled from the hardwood forest next to the campus.

Since my visit in 1993, the college has added an EcoVillage, an aesthetically pleasing complex of fifty apartments geared toward students with families, which uses 75 percent less water and energy than conventional housing and reuses at least 50 percent of all its wastes.

Racism and sexism have been fought at Berea, valiantly, since its founding in 1855. Back when Kentucky was still a slaveholding state, Berea's first freshman class had ninety-six blacks and ninety-one whites, and an almost equal number of women and men. Berea is named after a town in Acts 17:10 that was receptive to the Gospel, and the college is dedicated to the belief that "God has made of one blood all the peoples of the Earth."

Yes, it's Christian with a capital C, and no, the spoonbread at Boone's Tavern, the town's ancient restaurant, isn't to die for, but consider this: For about $250,000 you can get a three-bedroom house on a nice piece of land in one of the most seismically and volcanically stable regions in North America.

If any place is immune from Apocalypse 2012, it is Berea, Kentucky.

BE SAFE

Take 2012 seriously, but don't panic. Make contingency plans, but do nothing rash. We've got some work to do between now and then, lots of preparation, societal and personal, for the test that is to come. If we can find it in our hearts to look forward to it all, we'll also find a way to rise above the threat.

NOTES

My first job out of college in 1974 was as a library researcher, which basically entailed doing battle at the Mid-Manhattan Library on Fortieth Street and Fifth Avenue, across the street from its far more genteel, and infinitely slower, celebrity big brother, the New York Public Library. Getting unpleasantly high on the fumes from the chronically malfunctioning copy machines, or getting seasick from cranking through microfilm spools, was part of the job. Although I thank God for the Internet, which really burns out only one's eyes, I do miss the context and authority that the librarians gave to research back then. They knew their periodicals and could guide one toward quality information. Today, out on the Web, one must be one's own librarian.

Librarian duty has been particularly challenging with regard to researching this book, which after all ranges from hard science to oojie-boojie, from visionaries to kooks, across languages and continents. The sources below range from unimpeachable to idiosyncratic, as the subject demands.

A number of the citations below refer to the work of Tony Phillips, editor of Science@NASA (science.nasa.gov), an official NASA site for public information and education on space science. As editor, Phillips writes the articles that appear there, though in fact he is part of a production team that serves as a conduit for information and observations from NASA and from other sources that NASA considers reliable, such as the European Space Agency (ESA) and the Jet Propulsion Laboratory (JPL) in Pasadena, California. Phillips also performs this editorial service for a linked companion Web site, spaceweather.com.

INTRODUCTION

3 "I don't think the human race . . .": Highfield, "Colonies in Space," 12.

4 VIM-2 that breaks down antibiotics: Agence France Press, "Ultra Superbacteria."

5 "an entire zoo . . .": Kaku, "Escape from the Universe," 16.

5 hyperdense form of matter: Rees, *Our Final Hour*, 120–21, 123–25.

5 gray goo: Drexler, *Engines of Creation*, 171.

8 March 1989 solar radiation storm: Kappenman et al., "Geomagnetic Storms."

9 "11,000 years, at least": Solanki, 11/1/04.

10 California-sized cracks: Bentley, "Earth Loses Its Magnetism."

10 largest supervolcano: BBC2, "Supervolcanoes."

10 every 62 to 65 million years: Rohde and Muller, "Cycles in Fossil Diversity."

12 length of the lunar month: Sharer and Traxler, *The Ancient Mayan*, 116.

GUILTY OF APOCALYPSE: THE CASE AGAINST 2012

16 solar activity will next peak: Dikpati et al., "Unprecedented Forecast."

16 pole shift: Bentley, "Earth Loses Its Magnetism."

16 interstellar energy cloud: Dmitriev, "Planetophysical State."

16 the impact of a comet: Rohde and Muller, "Fossil Diversity."

17 most recent eruption: Smith and Siegel, *Windows into the Earth*.

17 Eastern philosophies: McKenna and McKenna, *Invisible Landscape*.

17 Armageddonist movement: Drosnin, *Bible Code*.

SECTION I: TIME

CHAPTER I

25 "At first glance . . . urban planning": Malmstrom, *Cycles of the Sun, Mysteries of the Moon*, 13.

27 environmental degradation: Diamond, *Collapse*.

27 "We have to wonder": Ibid., 177.

29 Milankovitch cycles: Locke, "Milankovitch Cycles."

30 cycle known as obliquity: Ibid.

31 "Prior to the 15th century": Barrios and Barrios Longfellow, *The Maya Cholqij*, 2.

32 As they write: Barrios and Barrios Longfellow, *The Maya Cholqij*, 4.

CHAPTER 2

40 center of an empire: Jenkins, *Maya Cosmogenesis*.

40 date is not 2012 but 2011: Calleman, *Mayan Calendar*.

SECTION II: EARTH

CHAPTER 3

53 poles last reversed: Bentley, "Earth Loses Its Magnetism."

53 "As to the changes . . . fern will grow": Hutton, "Small Pole Shift."

54 "In the first mechanism": Ibid.

54 "This type of mantle-slip . . . crustal tectonic movements": Ibid.

54 process may take a millennium: Bentley, "Earth Loses Its Magnetism."

55 a Danish satellite: Associated Press, "Report."

55 ozone controversy: Joseph, *Gaia*.

56 CFC destruction mechanism: Ibid.

CHAPTER 4

59 how Surtsey developed an ecosystem: Joseph, "Birth of an Island."

60 nuclear winter: Sagan and Turco, *Path Where No Man*.

60 enormous uranium reserves: Smith and Siegel, *Windows into the Earth*.

61 cover story: Wicks et al., "Uplift, Thermal Unrest."

61 Yellowstone supervolcano: BBC2, "Supervolcanoes."

62 "It would be": Ibid.

62 "I'm not sure": Ibid.

62 "columns or plumes . . . mantle and crust": Smith and Siegel, *Windows*.

63 "Of the roughly . . . Galapagos islands": Ibid., 29.

63 "The molten blobs . . . oceanic": Ibid., 28.

64 Marie's disease: BBC2, "Supervolcanoes."

65 new steam vents: LeBeau, "Letters."

65 "The only reasonable conclusion": Trombley, "Forecasting of the Eruption."

65 According to Steve Sparks: BBC2, "Supervolcanoes."

67 "restless" and "actively rising": Hill et al., "Restless Caldera."

67 "The most intense . . . beneath the Caldera": Ibid.

68 "There is evidence . . . intriguing question": Rymer, *Encyclopedia*.

70 "More recently, it has been realized": Irving and Steele, "Volcano Monitoring."

72 "was followed by at least": McGuire, *End of the World*, 104.

CHAPTER 5

77 Guatemalan dictator: Blythe, "Santiago Atitlán."

81 "As natural compensatory processes": Dmitriev, "Planetophysical State."

81 "hurricane refueling station": Kluger, "Global Warming."

81 may lead to earthquakes: Cowen, "Surprising Fallout."

SECTION III: THE SUN

CHAPTER 6

88 X2-class solar flare: Phillips, "X-Flare."

89 X7-class flare: Phillips, "Sickening Solar Flares."

90 "CMEs can account for": Phillips, "New Kind."

90 "most intense proton storm": Ibid.

91 "A bone marrow transplant": Ibid.

91 one of the most turbulent weeks: Ibid.

92 "Except possibly": Solanki, "Solar Variability."

93 jump from one steady state: Kluger, "Global Warming."

93 "the slow creep . . . controlled by a switch": Ibid.

94 "Global warming . . . is underway": Dougherty, "Lonnie Thompson."

95 "Tourism is the biggest": Ibid.

95 5,200 years ago: Thompson, "50,000-Year-Old Plant."

95 "Something happened . . . as well": Ibid.

96 "which means that the ice cap": Ibid.

98 expected to be turbulent: Phillips, "Solar Minimum."

98 largest solar flare: Whitehouse, "Explosion Upgraded."

100 "Energy balance . . . since 1850": Rottman and Calahan, "SORCE," 2.

101 "We predict": Dikpati and Gilman, "Unprecedented Forecast."

101 "When these sunspots decay": Ibid.

101 NCAR team's findings: Dikpati et al., "Predicting the Strength."

CHAPTER 7

102 Africa cracking apart: Associated Press, "New Ocean Forming."

102 "We believe . . . watching the phenomenon": Ibid.

104 "All Atlantic hurricanes . . . full-blown hurricanes": Sullivant, "Hurricane Formation."

107 far below average: Phillips, "Long Range Solar Forecast."

111 "The Earth's magnetic field": Pasichnyk, *Vital Vastness*, 869.

SECTION IV: SPACE

CHAPTER 8

119 "Increasing solar activity . . . is rather 'soft' ": Dmitriev, "Planetophysical State."

127 "This shock wave . . . to our Solar System": Ibid.

128 hydrodynamics of interplanetary plasma: Baranov, "Interstellar Medium."

128 Yellowstone-like geysers: Phillips, "Radical!"

129 "We've been monitoring": Phillips, "Jupiter's New Red Spot."

130 "The mission . . . to the Earth": Wolfe, "Alliance to Rescue Civilization," 2.

CHAPTER 9

133 "Cyanobacteria . . . in the sun": Margulis and Sagan, *Microcosmos,* 55.
134 uranium ore was oxidized: Joseph, *Gaia,* 193.
135 "Effects here on Earth . . . life on Earth": Dmitriev, "Planetophysical State."
136 "Since the Earth . . . the Earth's surface": Ibid.
138 "The biosphere . . . does not exist": Vernadsky, *Biosphere,* 44.
142 "At first . . . what I learned": Joseph, *Common Sense,* 141.
144 "The archeologists . . . blinds us to?": Argüelles, *Mayan Factor,* 20.
145 "The stars . . . of God": Votan, *Cosmic History,* 212.
146 how Yuri Knorozov: Coe, *Maya Lode,* 220–222.
149 "Important results . . . image of it": Kaznacheev and Trofimov, *Reflections on Life,* 38.

SECTION V: EXTINCTION

CHAPTER 10

155 led to the extinction: Sharpton, "Chicxulub," 7.
156 "bearded star": Leoni, *Nostradamus,* 175.
157 basic scientific knowledge: Lovelock, "Book for All Seasons."
158 will be killed: McKie, "Bad News."
159 mass extinctions: Rohde and Muller, "Cycles in Fossil Diversity."
159 "jumps out of the data": Kirchner and Weil, "Biodiversity."
159 key piece of evidence: Sharpton, "Chicxulub," 7.
162 comets have seeded Earth: Kellan, "Small Comets."
163 "NASA . . . water-bearing objects." Ibid.
164 Asteroid 1989 FC: Gerard and Barber, "Asteroids and Comets."
165 Shiva hypothesis: Rampino and Haggerty, " 'Shiva Hypothesis.' "
166 "A fiery dragon . . . his fellow men": Roads, "Mother Shipton's Complete Prophecy," 17.

SECTION VI: ARMAGEDDON

CHAPTER 11

175   reserves of helium-3: Joseph, "Mine the Moon."

175   divine code embedded: Drosnin, *Bible Code.*

176   scholarly article: Witztum, Rosenberg, and Rips, "Equidistant Letter Sequences."

176   "The rule is . . . until his end": Drosnin, *Bible Code,* 19.

177   Yitzhak Rabin: Ibid., 15.

178   comets are expected: Ibid., 155.

178   "Then I saw . . . Hebrew Armageddon": Ebor, *New English Bible.*

179   "While most Jews . . . unholy mess": Wells, "Unholy Mess."

179   battle would come: Lindsey, *Planet Earth.*

181   "He was dividing . . . leave it alone": Associated Press, "Divine Punishment."

184   "This interlocking . . . charge of it all": MacNeill, Winsemius, and Yakushiji, *Beyond Interdependence,* xxxii.

CHAPTER 12

193   "With the help of God": Gordon, "Kabbalist Urges Jews."

193   "The Mashiach . . . strict justice": Ibid.

194   "This declaration . . . our righteous Messiah.": Ibid.

195   "According to the writings . . . sabbatical year": Ibid.

196   "The Earth . . . humanity could transform": Levry, "Next 7 Years."

196   "Humanists stand . . . and women of light": Ibid.

CHAPTER 13

200   "The unique moment . . . interplanetary civilization": Argüelles, *Mayan Factor,* 194.

201   "Cosmic History": Votan, *Cosmic History,* 114.

202   "Human history": Argüelles, *Mayan Factor,* 9.

202   "Radiations from all stars . . . the biosphere": Vernadsky, *Biosphere,* 47.

203   a complex fractal: McKenna, "Temporal Resonance."

203 King Wen sequence: Ibid.

204 "For beauty is nothing": Rilke, *Duino Elegies*, "The First Elegy."

204 the 12/22/12 end point: McKenna and McKenna, *Invisible Landscape*.

206 "From the beginning . . . unity of its being": Griffiths, *Marriage of East and West*, 89, 92.

208 "The day Atlantis . . . went down": Geryl and Ratinckx, *Orion Prophecy*, 28.

209 "Maya insignia . . . air, earth": Janson, *Tikal*, 4.

210 "Four is the most": Hail, *Sacred Calendar*, 6.

210 "The Hopi prophecy . . . fifth world": Boylan, "Transition from Fourth."

210 "As we approach . . . new forms": Clow, *Catastrophobia*, 10.

216 "The time is coming . . . new age begin": Ebor, *New English Bible*.

217 "Lisa will . . . string theory": Kaku, "Escape from the Universe," 19.

218 "Spread across . . . was mathematics": Benford, "Mathematical Theology," 126.

CONCLUSION

226 using company funds: Landler, "Scandals."

228 op-ed piece: Joseph, "Who Will Mine."

230 psychologically whole: Jacobson, *Embattled Selves*.

230 mayor/administrator of a town: Ibid., 33.

232 "God has made": Peck and Smith, "First 125 Years," 26.

REFERENCES

.

Abraham, Ralph, Terence McKenna, and Rupert Sheldrake. *Trialogues at the Edge of the West: Chaos, Creativity, and the Resacralization of the World.* Rochester, VT: Bear & Co., 1992.

Agence France Presse. "Ultra Superbacteria Spreads to Asia." *Sydney Morning Herald,* July 18, 2002. smh.com.au/articles/2002/07/17/1026802710763.html.

Andersen, Peggy. "Mount St. Helens' Lava Baffles Scientists." Associated Press, January 2, 2006. news.yahoo.com/s/ap_on_sc/mount_st_Helens; _ylt=Ap53RM2yAENx.

Argüelles, José. *The Mayan Factor: Path Beyond Technology.* Rochester, VT: Bear & Co., 1996.

Associated Press. "New Ocean Forming in Africa." CBS News, December 10, 2005. cbsnews.com/stories/2005/12/10/tech/main/111579.shtml.

———. "Report: Earth's Magnetic Field Fading," CNN International, December 12, 2003. edition.cnn.com/2003/TECH/science/12/12/magnetic.poles.ap.

———. "Robertson: Sharon's Stroke Is Divine Punishment." *USA Today*, January 5, 2006. usatoday.com/news/nation/2006-01-05-robertson_x.htm.

Astronomical Applications Department, U.S. Naval Observatory, "The Seasons and the Earth's Orbit—Milankovitch Cycles," October 30, 2003. aa.usno.navy.mil/faq/docs/seasons_orbit.html.

Baranov, V. B. "Effect of the Interstellar Medium on the Heliosphere Structure" (in Russian). *Soros Educational Journal* no. 11, 1996, 73–79.

Barrios Kanek, Gerardo, and Mercedes Barrios Longfellow. *The Maya Cholqij: Gateway to Aligning with the Energies of the Earth.* Williamsburg, MA: Tz'ikin Abaj, 2004.

Benford, Gregory. "Applied Mathematical Theology." *Nature* 440 (March 2, 2006): BBC2 frompg. ff. 126.

Bentley, Molly. "Earth Loses Its Magnetism." BBC Online, December 31, 2003. news.bbc.co.uk/2/hi/science/nature/3359555.stm.

Blythe, Stephen. "Santiago Atitlán, Guatemala—A Human Rights Victory," 2004. gslis.utexas.edu/~gpasch/tesis/pages/Guatemala/otr04/hmnrts.htm.

Borchgrave, Arnaud de. "Later Than We think." *Washington Times,* February 6, 2006. washingtontimes.com/functions/print/print.pho?StoryID= 20060205-100341-6320r.

Boylan, Richard. "Transition from Fourth to Fifth World: The 'Thunder Beings' Return." Earth Mother Crying: Journal of Prophecies of Native Peoples Worldwide. wovoca.com/prophecy-rich-boylan-thunder-beings-htm.

BBC2. "Supervolcanoes." February 3, 2000. bbc.co.uk/science/horizon/ 1999/supervolcanoes_script.shtml.

Calleman, Carl Johan. *The Mayan Calendar and the Transformation of Consciousness.* Rochester, VT: Bear & Co., 2004.

Cameron, Alastair Graham Walte. "The Early History of the Sun." Smithsonian Miscellaneous Collections 151, no. 6 (July 15, 1966).

Clow, Barbara Hand, *Catastrophobia: The Truth Behind Earth Changes in the Coming Age of Light.* Rochester, VT: Bear & Co., 2001.

Coe, Michael D. *The Maya.* New York: Thames & Hudson, 1999.

Cowen, Robert C. "Global Warming's Surprising Fallout." *Christian Science Monitor,* August 19, 2004, p.16.

De Santillana, Giorgio, and Hertha Von Dechend. *Hamlet's Mill: An Essay Investigating the Origins of Human Knowledge and Its Transmission Through Myth.* Boston: Godine, 1977.

Diamond, Jared. *Collapse: How Societies Choose to Fail or Succeed.* New York: Viking, 2005.

Dikpati, Mausumi, Giuliana de Toma, and Peter A. Gilman. "Predicting the Strength of Solar Cycle 24 Using a Flux-Transport Dynamo-Based Tool." *Geophysical Review Letters* 33 L05102.doi:10.1029/2005/GL025221.

Dikpati, Mausumi, and Peter Gilman. "Scientists Issue Unprecedented Forecast of Next Sunspot Cycle." NCAR (National Center for Atmospheric Research) and UCAR (University Corporation for Atmospheric Research) Office of Programs, March 6, 2006. ucar/edu/news/releases/2006/sunspot/shtml.

Dmitriev, Alexey N. "Planetophysical State of the Earth and Life." *IICA Transactions* (1997), translated by A. N. Dmitriev, Andrew Tetenov, and Earl L. Crockett, Millennium Group, January 1, 1998. tmgnow.com/repository/global/planetophysical.html.

Dobson, Andrew, and Robin Carper. "Global Warming and Potential Changes in Host-Parasite and Disease-Vector Relationships." In *Global Warming and Biological Diversity*, edited by R. L. Peters and T. E. Lovejoy. New Haven: Yale University Press, 1994.

Dougherty, Maren. "Q&A with Glaciologist Lonnie Thompson." National Geographic Adventure Magazine, August 2004. nationalgeographic.com/adventure/0408/q_n_a.html.

Drexler, K. Eric. *Engines of Creation: The Coming Era of Nanotechnology.* New York: Anchor Books, 1986.

Drosnin, Michael. *The Bible Code.* New York: Touchstone/Simon & Schuster, 1997.

———. *Bible Code II: The Countdown.* New York: Viking/Penguin, 2002.

Ebor, Donald, ed. *The New English Bible with the Apocrypha.* New York: Oxford University Press, 1971.

Freidel, David, Linda Schele, and Joy Parker. *Maya Cosmos: Three Thousand Years on the Shaman's Path.* New York: William Morrow, 1993.

Gerard, Michael B., and Anna W. Barber. "Asteroids and Comets: U.S. and International Law and the Lowest Probability, Highest Consequence Risk." *New York University Environmental Law Journal* 6, no. 1 (1997): 3–40.

Geryl, Patrick, and Gino Ratinckx. *The Orion Prophecy: Will the World Be Destroyed in 2012?* Kempton, IL: Adventures Unlimited Press, 2001.

Golub, Leon, and Jay M. Pasachoff. *Nearest Star: The Surprising Science of Our Sun.* Cambridge: Harvard University Press, 2001.

Gordon, Baruch. "Kabbalist Urges Jews to Israel Ahead of Coming Disasters," October 17, 2005. Arutz Sheva, israelnn.com/news/php3?id=89850.

Griffiths, Bede. *The Marriage of East and West: A Sequel to the Golden String.* Springfield, IL: Templegate Publishers, 1982.

Griffiths, Bede. *A New Vision of Reality: Western Science, Eastern Mysticism and Christian Faith.* Springfield, IL: Templegate Publishers, 1990.

Hail, Raven. *The Cherokee Sacred Calendar: A Handbook of the Ancient Native American Tradition.* Rochester, VT: Destiny Books, 2000.

Hegel, G. W. F. *The Phenomenology of Mind.* Translated by J. B. Baillie. New York: Harper, 1967.

Highfield, Roger. "Colonies in Space May Be Only Hope, Says Hawking." *Daily Telegraph,* October 16, 2001, p. 12.

Hill, David P., et al. "Living with a Restless Caldera: Long Valley, California," USGS online, June 29, 2001. quake.wr.usgs.gov/prepare/factsheets/Long Valley/, Fact Sheet 108–96.

Howland, Jon. "Foes See U.S. Satellite Dependence as Vulnerable Asymmetric Target: Commercial Space Boom Comes with Risks, Absence of Public Debate Disturbing." *JINSA (Jewish Institute for National Security Affairs) Online,* December 4, 2003. globalsecurity.org/org/news/2003/031204-jinsa.htm.

Hutton, William. "A Small Pole Shift Can Produce Most, If Not All, of the Earth Changes Predicted in [Edgar] Cayce's Readings." *The Hutton Commentaries,* July 27, 2001. huttoncomentaries.com/PSResearch/Strain/SmallPoleShift.htm.

Irving, Tony, and Bill Steele. "Volcano Monitoring at Mount St. Helens: 1980–2005." Course description, University of Washington, July 2005. depts.Washington.edu/Chautauqua/2005/2005/Irving2.htm.

Jacobson, Kenneth. *Embattled Selves: An Investigation into the Nature of Identity through Oral Histories of Holocaust Survivors.* New York: Atlantic Monthly Press, 1994.

Janson, Thor. *Tikal: National Park Guatemala, A Visitor's Guide.* Antigua, Guatemala: Editorial Laura Lee, 1996.

Jenkins, John Major. *Maya Cosmogenesis 2012: The True Meaning of the Maya Calendar End-Date.* Rochester, VT: Bear & Co., 1998.

Joseph, Lawrence E. *Common Sense: Why It's No Longer Common.* Reading, MA: Addison-Wesley, 1994.

———. "Birth of an Island." *Islands,* May/June 1991, 112–15.

———. "Who Will Mine the Moon?" *New York Times,* January 19, 1995, p. A23.

———. *Gaia: The Growth of an Idea.* New York: St. Martin's Press, 1990.

Kaku, Michio. "Escape from the Universe: Wild, but Fun, Speculations from Physicist Michio Kaku." *Prospect* 107 (February 2005): 7–16.

Kappenman, John G., Lawrence J. Zanetti, and William A. Radasky. "Geomagnetic Storms Can Threaten Electric Power Grid." *Earth in Space* 9, no. 7 (March 1997): 9–11.

Kaznacheev, V. P., and A. V. Trofimov. *Cosmic Consciousness of Humanity: Problems of New Cosmogony.* Tomsk, Russia: Elendis-Progress, 1992.

———. *Reflections on Life and Intelligence on Planet Earth: Problems of Cosmo-Planetary Anthropoecology.* Los Gatos, CA: Academy for Future Science, 2004.

Kellan, Ann. "Scientist: Small Comets Bombard Earth Daily," May 28, 1997. CNN.com/TECH/9705/28/comet.storm/.

Kirchner, James W., and Anne Weil. "Biodiversity: Fossils Make Waves." *Nature* 434 (March 10, 2005): 147–48.

Kluger, Jeffrey. "Global Warming: The Culprit?" *Time,* October 3, 2005, pp. 42–46.

Kotze, P. B. "The Time-Varying Geomagnetic Field of Southern Africa." *Earth, Planets, Space* 55 (2003): 111–16.

Lagasse, Paul, ed. *The Columbia Encyclopedia,* 6th ed. New York: Columbia University Press, 2000.

Landler, Mark. "Scandals Raise Questions over Volkswagen's Governance." *New York Times,* July 7, 2005. nytimes.com/2005/07/07=business/world business/07Volkswagen.html.

LeBeau, Benny E. "Letters to the Spiritual Peoples of Mother Earth—November 17, 2003." Eastern Shoshone, Wind River Indian Reservation, January/February 2004. themessenger.info/archive/JanFeb2004/LeBeau.html.

Leoni, Edgar. *Nostradamus and His Prophecies.* Mineola, NY: Dover Publications, 2000.

Levry, Joseph Michael. "The Next 7 Years: The Heavenly Re-Positioning to Awaken Human Beings." *Rootlight,* 2004. rootlight.com/next7years intro.htm.

Lindsey, Hal. *The Late, Great Planet Earth.* Grand Rapids, MI: Zondervan, 1977.

Locke, W. W. "Milankovitch Cycles and Glaciation." Montana State Univer-

sity tutorial, spring 1999. homepage.Montana.edu/~geo1445/hyperglac/
time1/milankov.htm.

Lovelock, James. "A Book for All Seasons." *Science* 280, no. 5365 (May 8, 1998): 832–33.

Lovelock, James. *Gaia: A New Look at Life on Earth.* Oxford: Oxford University Press, 1979.

Malmstrom, Vincent H. *Cycles of the Sun, Mysteries of the Moon: The Calendar in Mesoamerican Civilization.* Austin: University of Texas Press, 1997.

———— "Izapa: Cultural Hearth of the Olmecs?" *Proceedings of the Association of American Geographers,* 1976, 32–35.

Margulis, Lynn, and Dorion Sagan. *Microcosmos: Four Billion Years of Microbial Evolution.* New York: Summit/Simon & Schuster, 1986.

McGuire, Bill. *A Guide to the End of the World.* New York: Oxford University Press, 2002.

McKenna, Terence. "Temporal Resonance." *ReVision* 10, no. 1 (Summer 1987): 25–30.

McKenna, Terence, and Dennis McKenna. *The Invisible Landscape: Mind, Hallucinogens, and the I Ching.* San Francisco: HarperSanFrancisco, 1993.

McKie, Robin. "Bad News—We Are Way Past Our 'Exinct by' Date." *The Guardian,* March 13, 2005. education.guardian.co.uk/higher/research/story/0,,1437163,00. html.

NOAA (National Oceanic and Atmospheric Administration). "Astronomical Theory of Climate Change," September 2002. ncdc.noaa.gov.

Pasichnyk, Richard Michael. *The Vital Vastness: Our Living Earth.* San Jose, CA: Writer's Showcase, 2002.

————. *The Vital Vastness: The Living Cosmos.* San Jose, CA: Writer's Showcase, 2002.

Peck, Elisabeth S., and Emily Ann Smith. *Berea's First 125 Years: 1855–1980.* Lexington: University Press of Kentucky, 1982.

Peterson, Scott. "Waiting for the Rapture in Iran." *Christian Science Monitor,* December 21, 2005. csmonitor.com.

Phillips, Tony. "Jupiter's New Red Spot," November 15, 2004. science.nasa.gov/headlines/y2006/02mar_redjr.htm? list47951.

————. "Long Range Solar Forecast: Solar Cycle Peaking Around 2022 Could Be One of the Weakest in Centuries," May 10, 2006.science.nasa.

gov/headlines/ y2006/10may_longrange.htm?list752889.

———. "A New Kind of Solar Storm," June 10, 2005. science.nasa.gov/head lines/y2005/10jun_newstorm.htm.

———. "The Rise and Fall of the Mayan Empire," November 15, 2004. science.nasa.gov/headlines/y2004/15nov_maya.htm.

———. "Sickening Solar Flares," January 27, 2005. science.nasa.gov/ headlines/y2005/27jan_solarflares.htm.

———. "Solar Minimum Explodes: Solar Minimum Is Looking Strangely Like Solar Max," September 15, 2005. science.nasa.gov/headlines/y2005/ 15sep_solarminexplodes.htm.

———. "X-Flare," January 3, 2005. spaceweather.com/index.cgi.

———. "Radical! Liquid Water on Enceladus," March 9, 2006. science.nasa .gov/headlines/2006/09/mar_enceladus.htm.

Pierce, Brian J. "Maya and Sacrament in Bede Griffiths," March 3, 2006. be-degriffiths.com/featured-article.html.

Posner, Richard A. *Catastrophe: Risk and Response.* New York: Oxford University Press, 2004.

Rampino, M. R., and B. M. Haggerty. "The 'Shiva Hypothesis': Impacts, mass extinctions, galaxy," abstract from article in *Earth, Moon, and Planets* 72 (1996): 441–60. pubs.giss.nas.gov/abstracts/1996/Rampino/Haggerty2.html.

Rees, Martin. *Our Final Hour: A Scientist's Warning: How Terror, Error and Environmental Disaster Threaten Humankind's Future in This Century—On Earth and Beyond.* New York: Basic Books/Perseus, 2003.

Rilke, Rainer Maria. "Duino Elegies," A. S. Kline's Poetry in Translation, August 18, 2006, www.tonykline.co.uk/PITBR/German/Rilke.htm# Toc509812215.

Rincon, Paul. "Experts Weigh Supervolcano Risks," BBC News, March 9, 2005. news.bbc.co.uk2hi/science/nature/4326987.stm.

Roach, John. "Stronger Solar Storms Predicted; Blackouts May Result," *National Geographic News,* March 8, 2006. nationalgeographic.com/news/ 2006/03/0306_060307_sunspots. html?source=rss.

Roads, Duncan. "Mother Shipton's Complete Prophecy." *Nexus Magazine* 2, no. 24 (February/March 1995): 17–21.

Rohde, Robert A. and Richard A. Muller. "Cycles in Fossil Diversity," *Nature* 434 (March 10, 2005): 208–10.

Rottman, Gary, and Robert Calahan. "SORCE: Solar Radiation and Climate Experiment." Laboratory for Atmospheric and Space Physics, (LASP), University of Colorado, and NASA Goddard Space Flight Center, 2004.

Russell, Peter. *Waking Up in Time: Finding Inner Peace in Times of Accelerating Change.* Novato, CA: Origin Press, 1992.

Rymer, Hazel. "Introduction." In *Encyclopedia of Volcanoes,* edited by Haraldur Sigurdsson et al. San Diego: Academic Press, 2000.

Sagan, Carl, and Richard Turco. *A Path Where No Man Thought: Nuclear Winter and the End of the Arms Race.* New York: Random House, 1990.

Schoonakker, Bonny. "Something Weird Is Going on Below Us: Satellites in Low-Earth Orbit over Southern Africa Are Already Showing Signs of Radiation Damage," *Johannesburg Sunday Times* (South Africa), July 18, 2004, p. 14.

Sharer, Robert J., with Loa P. Traxler. *The Ancient Maya,* 6th ed. Stanford, CA: Stanford University Press, 2006.

Sharpton, Virgil L. "Chicxulub Impact Crater Provides Clues to Earth's History." American Geophysical Union, *Earth in Space* 8, no. 4 (December 1995).

Sheldrake, Rupert. *A New Science of Life: The Hypothesis of Morphic Resonance.* New York: Jeremy P. Tarcher, 1981.

Smith, Robert B., and Lee J. Siegel. *Windows into the Earth: The Geologic Story of Yellowstone and Grand Teton National Parks.* New York: Oxford University Press, 2000.

Solanki, Sami K. "The Sun Is More Active Now than Over the Last 8,000 Years," Science Daily, November 1, 2004. sciencedaily.com/releases/2004/10/041030221144.htm.

Solanki, Sami K., and Natalie Krivova. "How Strongly Does the Sun Influence the Global Climate?" Max Planck Institute for the Advancement of Science, August 2, 2004. mpg.de/English/illustrationsDocumentation/documentation/pressReleases/2004/press.

———. "Solar Variability and Global Warming: A Statistical Comparison Since 1850." *Advanced Space Research* 34 (2004): 361–64.

Solara. *11:11: Inside the Doorway.* Eureka, MT: Star-Borne Unlimited, 1992.

Strong, Maurice. Introduction to *Beyond Interdependence: The Meshing of the World's Economy and the Earth's Ecology,* eds. Jim MacNeil, Peter Winsemius, and Taizo Yakushiji. New York: Oxford University Press, 1991.

Sullivant, Rosemary. "Researchers Explore Mystery of Hurricane Formation," *Earth Observatory,* September 23, 2005. jpl.nasa.gov/news/features .cfm?feature=942.

Thompson, Lonnie. "50,000-Year-Old Plant May Warn of the Death of Tropical Ice Caps." *Research News,* Ohio State University, December 15, 2004. osu.edu/archive/quelplant.htm.

Trombley, R. B. "Is the Forecasting of the Eruption of the Yellowstone Supervolcano Possible?" Southwest Volcano Research Centre, 2002. getcited.org .publ/103379403.

Unger, Craig. "Apocalypse Soon!" *Vanity Fair,* December 2005, pp. 204–22.

Upgren, Arthur R. *Many Skies: Alternative Histories of the Sun, Moon, Planets, and Stars.* New Brunswick, NJ: Rutgers University Press, 2005.

Vernadsky, Vladimir I. *The Biosphere.* Translated by David B. Langmuir. New York: Copernicus/Springer-Verlag, 1998.

Votan, Pakal, and Red Queen. *Cosmic History Chronicles: vol. 1, Book of the Throne: The Law of Time and the Reformulation of the Human Mind.* Watertown, NY: Foundation for the Law of Time, 2005.

Ward, Charles A. *Oracles of Nostradamus.* Whitefish, MT: Kessinger Publishing, 2003.

Wells, Jeff. "Unholy Mess Brewing on the Temple Mount." Rigorous Intuition, January 17, 2005. rigorousintuition.blogspot.com/2005/01/unholy-mess-brewing-on-the-temple-mount.html.

Whitehouse, David. "Sun's Massive Explosion Upgraded," BBC News, March 17, 2004. news.bbc.co.uk/2/hi/science/nature/3515788.stm.

Wicks, Charles W., et al. "Uplift, Thermal Unrest and Magma Intrusion at Yellowstone Caldera." *Nature* 440 (March 2, 2006): 72–75.

Wilcox, Joan Parisi. *Keepers of the Ancient Knowledge: The Mystical World of the Q'ero Indians of Peru.* Boston: Element Books, 1999.

Wilhelm, Richard, and Cary F. Baynes, trans. *The I Ching, or Book of Changes.* Princeton: Princeton University Press, 1967.

Witztum, Doron, Eliyahu Rips, and Yoav Rosenberg. "Equidistant Letter Sequences in the Book of Genesis." *Statistical Science,* vol. 9, no. 3, 1994, 429–38.

Wolfe, Steven M. "The Alliance to Rescue Civilization: An Organizational Framework." Paper presented at the Space Frontier Foundation, Fourth Annual Return to the Moon Conference, Houston, TX, July 20, 2002.

INDEX

Abraham, 171–72, 189

Abraham, Ralph, 205

Abu Jahal, 173–75, 188

Aerospace Consulting Corporation (AC2), 2, 3

African fissure, 102–4

Ahmadinejad, Mahmoud, 185, 186, 191

Akademgorodok, Russia, 124–25

Al Aqsa mosque, 189, 192, 195

Al-Jazzar, 219–20

Alley, Richard, 93

Alpha Centauri star system, 131

Al Qaeda, 66

Alvarez, Luis, 155, 159–60

Anaerobes, 133

Annan, Kofi, 183

Apocalypse 2012, 1–2

  allure of doomsday, 213–14

  Argüelles's predictions, 199–200, 201–2

  Armageddon movement and, 181, 184

  Atlantis and, 208–9

  Bible code and, 178

  "deadline"perspective on, 14–15

  divorce and, 42

  Dmitriev's views on, 142–43

  "doomsday chic"phenomenon, 212–13

  doomsday cults and, 215

  faith in something coming next, human need for, 215–17

  fears regarding, 49, 214–15

  "glorious new age"alternative to Apocalypse, 74–75

  humanity's options for response to, 222–32

  I Ching and, 203–4

  Joseph's disclaimers about, 13–14

  Kalki's predictions, 207–8

  Levry's views on, 196–97

man-made catastrophe, potential for, 122

Maori beliefs about, 211

Native American beliefs about, 210

population explosion and, 154

prayer and, 224–25

prevention and relief efforts regarding, 228–29

as projective test for anyone who contemplates it, 75

proliferation of 2012 predictions, 211–12

psychological defenses against, 230

solar activity and, 100–101, 114

space-time method for avoidance of, 144–46

spirit world as source of information about, 141–42

subculture concerned with, 11–12

See also Mayan prophecies for 2012

"Applied Mathematical Theology"(Benford), 218

Arafat, Yassir, 181

Archeologists, 39–41

Argüelles, José, 144, 145, 147, 148, 149, 199–200, 201–2

Armageddon movement

Apocalypse 2012 and, 181, 184

divine retribution and, 181–82

global warming and, 196

Jews of Israel and, 192–96

messianic beliefs and, 191–96

overview of, 178–81

resisting the prophecy, 197–98

Temple Mount and, 189–90, 192

world government and, 182–84

Armstrong, Herbert W. and Garner Ted, 192

Asteroids, 163–64. *See also* Extraterrestrial impacts

Astrology, 110

Astronomy

of Mayans, 12, 28, 31, 37, 208

mythology and, 200–201

of Sumerians, 160–61

Atitlán lake, 73

Atlantis, 208–9

Atom smashers, 4–5

Auroras, 52, 53, 128

Austria, Holocaust and, 185–86

Axis tilt of Earth, 30

Ayalew, Dereje, 102–3

Baranov, Vladimir B., 128, 129

Barrios, Carlos, 34–36, 37, 40–42, 57, 145–46, 199, 201, 202, 231

Barrios, Gerardo Kanek, 24, 32, 36–37, 38, 40, 41, 42, 107, 199, 201, 202, 209–10, 231

Bast, Robert, 103

Begin, Menachem, 181

Benford, Gregory, 218

Berea, Kentucky, 231–32

Berger, Oscar, 80

Bermuda Triangle, 103

Bible code, 17, 175–78, 187

Bible Code, The (Drosnin), 175–76, 177–78

Big bang, 217–18

Bilderberg organization, 182–83

Biochemical weapons, 3–4

Biosphere, 135–36, 137–38

galactic radiation and, 202

Biosphere, The (Vernadsky), 138

Black holes, 4–5, 13, 33

Blair, Tony, 182

Bloxham, Jeremy, 54
Bolides, 164
Bonaparte, Napoleon, 186, 219–20
Bono, Tony, 34
Boylan, Richard, 210
Brundtland, Gro Harlem, 183–84
Burgess, Thomas, 113, 114
Bush, George W., 178, 198

Calendar system of Mayans, 12, 24–25,
    144
Calleman, Carl Johann, 40
Cayce, Edgar, 53, 54
Chaim, Rabbi Yosef, 193
Chanel, Coco, 56
Cherokee Sacred Calendar, The (Hail),
    210
Chicxulub impact, 155, 156, 159
Chlorofluorocarbons (CFCs), 55–57
Christiansen, Robert, 61
Cicerone, Ralph, 56
Cilliers, Pierre, 215–17
Climate change
    climate catastrophe of 5,200 years
        ago, 95
    scientific research on, 93–95,
        96–98
    Sun-Earth relationship and, 29–30,
        92–93, 100, 108–9
    See also Global warming
Clinton, Bill, 182
Cloud formation, 51
Clow, Barbara Hand, 210–11
Coe, Michael, 146
Collapse (Diamond), 27
Colonization of space, 130–31, 227
Comets, 156, 161–63. See also
    Extraterrestrial impacts

Compendium of Fossil Marine Animal
    Genera (Sepkoski), 158
Core, The (film), 52
Cosmic History Chronicles (Argüelles),
    201
Cosmic rays, 96–98
Creation, ideas about, 75, 76
Crutzen, Paul, 56
Cullman, Brian, 212
Cyanobacteria, 132–33
"Cycles in Fossil Diversity"(Muller and
    Rohde), 158, 159

Dante, 173–74
David, King, 189
Day After Tomorrow, The (film), 94
Death, randomness of, 157
Dechend, Hertha von, 200
Deep Impact space probe, 163
Diamond, Jared, 27
Dikpati, Mausumi, 101
Dinosaurs, extinction of, 155, 156
Dmitriev, Alexey, 81, 117–18, 119–20,
    122, 123–24, 125, 126, 127,
    129–30, 135, 136, 142–43, 146,
    161, 202
Dome of the Rock, 189
Drexler, Eric, 6
Drosnin, Michael, 175–76, 177–78
Duino Elegies (Rilke), 204

Earth Mother Crying (Boylan), 210
Earthquakes
    African fissure and, 102–4
    megacatastrophes, 80–82, 105
    number of, 69
    preparedness for, 71
    supervolcanoes and, 67

Earth Summit of 1992, 183
Egyptian influence on Mayan culture, 209–10
Einstein, Albert, 145
Electricity in atmosphere, 136–37
El Greco, 123
Eliyahu of Vilna, Rabbi, 176–77, 195
Embattled Selves (Jacobson), 230
Engines of Creation (Drexler), 6
Environmentalism, 183, 184
"Equidistant Letter Sequences in the Book of Genesis"(Witztum et al.), 176–77
Ether, 41
Evangelical Christians, 180, 192
Extinctions, 10
  extraterrestrial impacts and, 155, 156, 159, 161, 165–66
  mass extinctions, 158–61, 165–66
  population explosions and, 153–54
  supervolcanoes and, 63–64
Extraterrestrial impacts, 16
  Bible code and, 178
  defense systems against, 164–65
  extinctions and, 155, 156, 159, 161, 165–66
  global warming and, 165
  Shiva hypothesis, 165–66
  sources of comets and asteroids, 161–62
  volcanism and, 165

Federal Emergency Management Administration (FEMA), 66–67
Feynman, Richard, 100
Finnboggadottir, Vigdis, 58
Foster, Jodie, 56

Foundation for the Law of Time, 147
Frank, Louis, 162–63

Gabriel, Archangel, 171, 172, 173, 174, 175
Gagarin, Yuri, 120, 121
Gaia hypothesis, 68, 123, 135, 175, 212, 224
Galactic radiation, 202
Galilee World Heritage Park, 180
Garcia, Gen. Fernando Romeo Lucas, 77
Gates, Melinda, 182
Genesis, Book of, 76, 176–77
Germany, Holocaust and, 185–86
Geryl, Patrick, 208–9
Gilman, Peter, 101
Glacial melting, 95–96
Gladwell, Malcolm, 72
Global warming, 229
  Armageddon movement and, 196
  extraterrestrial impacts and, 165
  glacial melting and, 95–96
  megacatastrophes and, 81–82
  volcanism and, 68–70, 71–72
Gore, Al, 94, 178, 183
Gould, Stephen Jay, 36
Gravity, 112–13
Gray goo, 5–6
Griffiths, Father Bede, 206
Guide to the End of the World, A (McGuire), 72
G'Zell, Otter, 212

Haggerty, B. M., 165
Hail, Raven, 210
Haile Selassie, emperor of Ethiopia, 102

Hameed, Sultan, 108–9
Hamlet's Mill (Santillana and
    Dechend), 200
Hathaway, David, 107
Hawking, Stephen, 3
Heliosphere, 126–27
Hipparchus, 28
Holland, Heinrich, 133
Holocaust controversies, 184–86
Homeostasis, 135
Hot spots, 62–63, 64
Hurricanes
    Atlantic hurricanes, origin of, 104–5
    destruction from, 79–80
    megacatastrophes, 80–82, 105
    Sun-Earth relationship and, 99–100
Hurricane Stan, 79–80
Hutton, William, 53–54
Huxley, Aldous, 73

Ice ages, 92–93, 108–9
I Ching (Book of Changes), 203–4
International Heliophysical Year (IHY)
    2007, 105–6
Interplanetary magnetic field (IMF), 51
Interstellar energy cloud, 9–10, 16, 161
    colonization of space and, 130–31
    data regarding, 127–28
    Earth, possible effect on, 135–37
    heliosphere shock wave and, 126–27
    planets, impact on, 128–29
    solar activity and, 119–20, 129–30
Interstellar space, heterogeneity of,
    125–26
Invisible Landscape, The (McKenna), 204
Io, 130
Iridium, 159
Irving, Tony, 70

Ishmael, 172
Island formation, 58–59
Izapa ruin, 40

Jacobson, Kenneth, 230
Janson, Thor, 209
Jenkins, John Major, 40
Jerusalem, 189–90
Jesus Christ, 190, 216–17
Jewish Defense League, 194
Jews of Israel, 192–96
John the Baptist, 123
Joseph, Edward D., 121–22, 157
Josephson, Brian, 205
Jung, Carl, 203
Jupiter, 128–29, 162

Kaaba shrine, 172–73
Kaduri, Rabbi Yitzhak, 192–93, 194,
    195, 196
Kael, Pauline, 157
Kahane, Meir, 194–95
Kahn, Herman, 9
Kaku, Michio, 5
Kalki Bhagavan, Sri, 207–8
Kaznacheev, V. P., 149
Keepers of the Ancient Knowledge
    (Wilcox), 210
Kermode, Frank, 213
Khalid (Sword of Allah), 188
Kilimanjaro, 94–95
Kirchner, James, 159
Kirtland Air Force Base, 2–3
Kissinger, Henry, 182
Knorozov, Yuri, 146–47
Kotze, Pieter, 50, 51, 52, 54, 55
Kozyrev, Nikolai A., 143, 146
Kozyrev mirror, 143

Kristofersson, Capt. Eirikur, 141–42

Kuiper belt, 162

Kuru, 4

Kuznetsova, Taisia, 147–48

Kyoto Protocol, 183

LaHaye, Tim, 182

Landa, Father Diego de, 32

Large hadron collider (LHC), 5

Late Great Planet Earth, The (Lindsey), 179

Leary, Timothy, 203

LeBeau, Bennie, 64–65

Left Behind series, 182

Levry, Joseph Michael, 196–97

Lin, Robert, 90

Lindsey, Hal, 179

LISA (laser interferometer space antenna), 217–18

Llamatepec volcano, 80, 81

Longfellow, Mercedes Barrios, 24

Long Valley supervolcano, 67

Loon, Harry van, 89, 100

Lord Byron, 73, 74, 76, 79

Lovelock, James, 55, 68, 123, 136, 138, 157, 212

Magnetic field of Earth, 10, 16, 47
  cosmic rays and, 97–98
  cracking apart of, 55
  depletion of, 50, 51, 52, 54, 55, 57
  origins of, 50, 51
  ozone depletion and, 55–57
  pole reversal, 52–54
  predictions about, 53
  psychic communication and, 143–44

Sun-Earth relationship and, 50–51, 54, 91, 111

Van Allen radiation belts and, 51–52

Malmstrom, Vincent H., 25–26

Mantle-slip mechanism, 54

Maori people, 211

Maran, Steve, 163

Margulis, Lynn, 68, 133, 134, 212

Mars, 129, 130

Marx, Karl, 179

Maximón, the playboy saint, 74

Maya, The (Coe), 146

Maya Cholqij, The (Barrios and Longfellow), 24, 32

Maya Cosmogenesis 2012 (Jenkins), 40

Mayan culture
  archeologists and, 39–41
  Argüelles's connection to, 199–200, 201–2
  astronomy of, 12, 28, 31, 37, 208
  ball game, 39
  calendar system, 12, 24–25, 144
  cataclysmic prophecies, possible reasons for, 155
  codices, 32, 38, 146–47
  counting system, 24–25
  creation story, 75
  degeneration and collapse of classic Mayan civilization, 26–27
  ecological partnership, spirit of, 75–76
  Egyptian influence on, 209–10
  funeral protocols, 79–80
  Guatemalan civil war and, 77–78
  hieroglyphics, 147
  Izapa ruin, 40
  Maximón, the playboy saint, 74

personal connection with cosmos, sense of, 78

Russian-Mayan intellectual collaboration, 146–47

sacrifice, tradition of, 39

shaman's life, 34–36, 37, 41

sky-watching as key to predicting the future, 31–32

Tikal ruins, 23, 28

time, concepts of, 25–26

Western scholarship on, 32

wheel technology, 40

See also Mayan prophecies for 2012

Mayan Factor, The (Argüelles), 144, 201

Mayan prophecies for 2012, 16, 187

Barrios brothers' views on, 34–37, 40–42, 231

calendar system and, 24–25

"divine origin" of, 30–31

eclipse of Earth's view of center of Milky Way, 13, 32–33

as mass of factoids, 23–24

"new age" to begin 12/21/12, 12–13

Resurrection of ancestors, 36, 74

as revenge hoax on the North, 38–39

2011 alternative, 40–41

McGuire, Bill, 72

McKenna, Terence, 203–5

Mead, Margaret, 213

Megacatastrophes, 80–82, 105

Mendoza Mendoza, Juan Manuel, 74, 75, 76, 78–79, 80, 82

Messianic beliefs, 191–96

Metallic ore deposits, energy-conducting effect of, 136–37

Milankovitch, Milutin, 29

Milankovitch cycles, 29–30

Milky Way, eclipse of Earth's view of center of, 13, 32–33

Mithraism, 29

Molina, Mario, 56

Moon, 130, 174, 175

Morissette, Alanis, 48

Mormons, 174

Mosley-Thompson, Ellen, 94

Mount Rainier volcano, 70

Mount St. Helens volcano, 62, 69, 70

Muhammad, Prophet, 172–75, 189, 191

Muller, Richard, 158, 159, 160, 161, 165

Mythology, 200–201

Nanotechnology, 5–6

Napoleon's sword, 219–21, 223–24

National Center for Atmospheric Research (NCAR), 101

Native American beliefs about Apocalypse 2012, 210

Nature as universal mind, 205–6

Nemesis hypothesis, 160

Neptune, 128

Netanyahu, Benjamin, 194

New Science of Life, A (Sheldrake), 205

Newton, Isaac, 112, 176

Nixon, Richard, 169

Noosphere, 138–41

Nostradamus, 156

Nuclear fusion research, 228

Nuclear reactors, natural, 134

Nuclear winter, 60, 71–72

Oneness concept, 206
123Alert organization, 47, 70
Oort, Jan Hedrick, 161
Oort cloud, 161–62, 165
Opik, Ernst, 161
Orion Prophecy, The (Geryl and
    Ratinckx), 208–9
Orton, Glenn, 129
Oval BA storm, 129
Oxygen in atmosphere, 133
Ozone depletion, 55–57

Pasichnyk, Richard Michael, 111
Penzias, Arno, 117
Perseid meteor shower, 156
Phillips, Tony, 8, 91, 98
Planetary configurations, 109–13
"Planetophysical State of the Earth and
    Life"(Dmitriev), 127
Planet X, 128, 160–61
Plato, 186, 208
Pluto, 128
Population explosions, 153–54
Prayer, 224–25
Precession process, 28–29
Prions, 4
Proton storms, 89–91
Psychic phenomena, 124
    experiments in, 147–49
    magnetic field of Earth and,
        143–44
    noosphere and, 138–41
    spirit guides, 141–42
Pynchon, Thomas, 144
Pythagoras, 31

Quayle, Dan, 164
Quelccaya ice cap, 96

Quran, 173, 174
Quraysh tribe, 172–73

Rabin, Yitzhak, 177
Radioactive isotope dating techniques,
    96–98
Rain forests, 135–36
Rampino, M. R., 165
Rao, Joe, 92
Ratinckx, Gino, 208–9
Razilo, Ramil, 180
Reagan, Ronald, 192
Rees, Martin, 5
Relacion de las cosas de Yucatán (Landa),
    32
Remy, Roger, 7
RHESSI satellite, 107
Rickover, Adm. Hyman, 11
Rilke, Rainer Maria, 204
Rips, Eliyahu, 176–77
Robbins, Tom, 204
Robertson, Pat, 181–82, 190
Rock, Chris, 49
Rohan, Michael Dennis, 192
Rohde, Robert, 158, 159, 160, 165
Rosenberg, Yoav, 176–77
Rother, Father Stanley
    "Francisco,"76–77
Rowland, F. Sherwood (Sherry), 56
Rushdie, Salman, 174
Rymer, Hazel, 68

Sagan, Carl, 60
Sagan, Dorion, 133
Sandia National Laboratories, 3
Sanhedrin religious tribunal, 194–95
Santillana, Giorgio de, 200
Santorini supervolcano, 65

Satanic Verses, 174–75
Satellites for solar research, 106–7
Saturn, 128
Schneerson, Rabbi Menachem, 193
Schwarzenegger, Arnold, 225
Sepkoski, Jack, 158–59
Shampoo (film), 157
Sharks, 47, 49
Sharon, Ariel, 181–82, 190
Shehab, 188
Shehab, L'Emir Bashir, II, 219–20,
    221
Shehab, Kamil, 220
Sheldrake, Rupert, 205–6
Shipton, Mother, 165–66
Shiva hypothesis, 165–66
Shoemaker-Levy 9 comet, 162
Siegel, Lee J., 62, 63
Smith, Robert B., 61–62, 63
SOHO satellite, 106
Solanki, Sami, 92, 93, 99
Solar Radiation and Climate
    Experiment (SORCE), 87–88,
    97–98, 99, 100
Solar System
    center of mass of, 113–14
    heating up of, 124
    as organismic entity, 163
Solomon, King, 189
SORCE satellite, 107
Soros, George, 128
South, Stephanie, 201
Space exploration, 91, 130–31, 227
Space-time, 144–46
Sparks, Steve, 62, 65
Spirit guides, 141–42
Sport-utility vehicles (SUVs), 225–26
Stander, Anne, 70–71, 140

Star Trek (TV show), 45, 54
Steele, Bill, 70
STEREO satellite, 107
Stolarski, Richard, 56
Strangelets, 5
String theory, 145
Stromboli volcano, 69
Strong, Hanne, 183
Strong, Maurice, 182–83, 184
Sumerian astronomers, 160–61
Sun, 16
    Apocalypse 2012 and, 100–101, 114
    binary companion, possible, 160
    changes within, 9
    coronal mass ejections (CMEs),
        89–91, 99
    exceptionally energetic behavior of
        recent years, 92, 93
    Halloween storms of 2003, 98–99
    historical solar activity, techniques
        for determining, 96–98
    imprisoned radiation, release of, 114
    interstellar energy cloud and,
        119–20, 129–30
    Maunder Minimum of 1645 to 1715,
        108–9
    Mayan laws protecting, 78
    planetary configurations, impact of,
        109–13
    predictions regarding solar activity,
        100–101
    scientific research regarding, 105–7
    solar flares, 88, 89–90, 91, 98–99
    sunspot cycle, 88–89
    sunspots, 7, 8–9, 88, 96–97, 98, 101,
        108
    2005 solar activity, 88, 89–92, 98, 99,
        102, 103

See also Solar System; Sun-Earth
  relationship
Sun-Earth relationship, 16, 87–88
  climate change and, 29–30, 92–93,
    100, 108–9
  cosmic rays and, 96–98
  cracking of Earth's crust and, 102–4
  hurricanes and, 99–100
  imbalance as threat to, 107
  magnetic field of Earth and, 50–51,
    54, 91, 111
  Milankovitch cycles and, 29–30
  ozone depletion and, 56
  proton storms and, 89–91
  social chaos and, 108–9
  space exploration and, 91
  two-way energy relationship,
    109–11
Supervolcanoes, 10, 17, 59
  calderas of, 61
  characteristics of, 62
  earthquakes and, 67
  evidence of impending eruption,
    61–62, 64–65
  explosive force of, 62
  extinctions related to, 63–64
  global warming and, 71–72
  government silence on dangers of,
    64–65
  historical record of, 60–61, 65, 71–72
  hot spots and, 62–63, 64
  magma dynamics of, 65
  nuclear-winter-type catastrophe,
    potential for, 60, 71–72
  odds of supervolcano eruption in
    our lifetime, 67–68
  periodicity of eruptions, 61

  prevention of eruptions, 65–66
  responsive measures regarding,
    66–67
  terrorism and, 66
Surtsey island, 58–59
Swift-Tuttle comet, 156
Swimme, Brian, 201, 202

Tanning, 56
Teilhard de Chardin, Pierre, 138
Tempel 1 comet, 163
Temple Mount, 189–90, 192, 195
Terrorism, 66
Thompson, Lonnie, 93–96
Tikal ruins, 23, 28
Time
  arrows and cycles of, 26
  Mayan concepts of, 25–26
  space-time, 144–46
Time travel, 143, 146
Toba supervolcano, 71–72
Toma, Giuliana de, 101
TRACE satellite, 107
Trialogues at the Edge of the West
    (McKenna et al.), 205
Trofimov, Alexander V., 143, 144,
    147–49
Trombley, R. B., 65
Tsunami of 2004, 7–8
Turner, Ted, 183
2012 Apocalypse. See Apocalypse 2012

Ultraviolet radiation, 56
United Nations, 182, 183, 184
United States Geological Survey
    (USGS), 67
Uranus, 128

V. (Pynchon), 144

Van Allen, James A., 51

Van Allen radiation belts, 51–52

Van Gogh, Vincent, 31

Venus, 129, 207–8

Vernadsky, V. I., 138, 202

VIM-2 enzyme, 3–4

Vital Vastness, The (book), 111

Volcanism

    extraterrestrial impacts and, 165

    global warming and, 68–70, 71–72

    island formation and, 58–59

    megacatastrophes, 80–82

    psychic's predictions regarding, 70–71

    total amount of volcanic activity, 68–69

    See also Supervolcanoes

Volkswagen Phaeton automobile, 226

Voluntary Human Extinction Movement, 131

Voorhies, Michael, 63–64

Vulcan Plasma Disintegrator, 2, 4, 7

Water on Earth, origin of, 162–63

Weissmandel, H. M. D., 176

Wells, Jeff, 179

Wiesenthal, Simon, 185

Wilcox, Joan Parisi, 210

Wilson, Robert, 117

Windows into the Earth (Smith and Siegel), 62, 63

Witztum, Doron, 176–77

Wolfe, Steven M., 130

World government, 182–84

World Health Organization (WHO), 184

Worldwide Church of God, 192

Yellowstone supervolcano, 10, 59, 60–62, 63–67

Yokoh Satellite B, 107

Ziegler, Ron, 169

ABOUT THE AUTHOR

Lawrence E. Joseph is a journalist and science consultant who has written extensively on scientific matters, the environment, politics, and the business world for publications including the *New York Times*, *Salon*, and *Audubon*. He is currently chairman of the board of Aerospace Consulting Corporation in Albuquerque, New Mexico.